Practical Boundary Surveying

Paul Gay

Practical Boundary Surveying

Legal and Technical Principles

Springer

Paul Gay
Westport, MA
USA

ISBN 978-3-319-07157-2 ISBN 978-3-319-07158-9 (eBook)
DOI 10.1007/978-3-319-07158-9

Library of Congress Control Number: 2014945258

Springer Cham Heidelberg New York Dordrecht London

Printed on acid-free paper

Springer is part of Springer Science+Business Media (www.springer.com)

Disclaimer

This book is intended to present the reader with an overview of general principles of land surveying and the general legal principles underlying boundary establishment and easements. Because statutes and common law vary from jurisdiction to jurisdiction the information presented herein may not be applicable in the reader's jurisdiction or to any particular circumstances in any jurisdiction. Because the conclusions and observations presented in this book rely on the author's opinion, the information presented herein is subject to interpretation by other professionals in the field who may be more familiar with laws, precedent and practices in their own jurisdiction, and they may have their own opinion on the subjects presented in this book. Prior to relying on the materials presented herein, the reader is encouraged to research laws, methods and requirements of land surveying existing in the reader's jurisdiction and survey system and to seek land surveying advice from a land surveyor licensed in the jurisdiction and legal advice from an attorney licensed to practice in that jurisdiction. This book is not, under any circumstances, to be construed as providing legal advice or surveying advice and the author shall not, under any circumstances, be liable for losses by readers who rely on any information presented herein.

To my wife Viola, a tireless researcher and a surveyor in her own right, and to my mother Elinor who encouraged me to become a surveyor.

Contents

About the Author

Paul Gay is an attorney and a professional land surveyor. He has been a land surveyor in private practice for more than 35 years and has been primarily involved in boundary surveying with a special interest in boundary dispute resolution. Mr. Gay has served as an expert witness in many trials involving land and boundary issues from Massachusetts and Rhode Island to Alaska. He wrote the book *Fundamentals of Boundary Surveying*, Professional Surveyor Publishing Co., Inc. in 2002, *Survey*, a computer program for surveying calculations, *Metrology*, a Windows units conversion program, and *Tide*, a program for calculating tides for any date at many U.S. coastal locations. Other publications include articles in surveying publications, *A History of Gray's Mill, Basic Principles of 19th Century Water Power* and *Sediment Transport around Gooseberry Neck.* Mr. Gay holds a B.S. degree in Sociology from the University of Massachusetts, an ScM degree in Geological Sciences from Brown University and a J.D. degree from New England Law.

Chapter 1
Introduction

This is a book about boundary surveying. It is written for people who own land and for people who work in any profession or job which involves land. It is also written for land surveyors and students of land surveying. All parcels of land have boundaries. For many parcels of land, the location of the boundaries is uncertain, unknown or disputed. This book will provide the reader with a background on boundary surveying techniques and some of the common legal issues which govern boundary establishment.

Boundary surveying is sometimes thought to be primarily an engineering discipline. In reality, there is much more to boundary surveying than engineering. For most boundary surveyors, that is what makes the profession so challenging and interesting. You can't learn to be a competent surveyor by taking a course, acquiring a degree or reading a book—although all of these things help to provide the necessary foundation. Boundary surveying combines science, engineering, mathematics and law. Correctly performing boundary surveys also requires years of experience. Some boundaries are relatively easy to establish. More often than not they are difficult to establish. Surveyors routinely measure miles within fractions of an inch. Yet even with this ability, they are sometimes unable to determine the actual extent of ownership within several feet. In some surveys, the uncertainty is even greater. How can this be? This book will help the reader to understand why such uncertainties exist and what remedies may be available.

One purpose of this book is to acquaint people who are not land surveyors with the principles used by land surveyors to establish property lines. Home owners, real estate agents, attorneys, engineers, city planners, building officials, students, bankers, title researchers, GIS practitioners and many others will find the information in this book indispensable when they have questions relating to boundaries, easements and land surveying in general.

This book will also be useful to land surveyors who want or need to review the legal principles of boundary establishment and easements. It will serve as an

P. Gay, *Practical Boundary Surveying*, DOI 10.1007/978-3-319-07158-9_1

important learning resource for students who are studying land surveying at the college level and who hope to become licensed. Because this book focuses on boundaries, it is not intended to replace a general textbook covering all types of surveying, such as construction and topographic surveying. However, because it integrates legal, technical and operational aspects of boundary surveying it will help a student develop an overall view of how surveys are performed.

The technical methods and equipment used in boundary surveying have changed as a result of the widespread adoption of computers and other electronics. Total stations (electronic theodolites) and global positioning systems (GPS) have replaced transits and steel tapes. Data collectors have replaced field books. Computers have replaced calculators for most surveying calculations. Large format plotters have replaced drawing plans by hand. However, these changes have no effect on old boundary surveys which have endured for generations. Stone boundary markers and piles of stones set in the 1800s are still routinely found in fields and in the woods. Plans and deeds created and recorded in the same period are still found in recorder's offices throughout the country. Boundary surveying is inextricably tied to history. Surveyors and others, who need to understand how a boundary originated and how it should be reestablished today, must be familiar the methods and equipment that were used when the boundary was first created. With some modern texts on land surveying, the emphasis is on the new, the latest equipment and techniques that are being used today. Although there is nothing wrong with this approach, indeed surveyors must understand and be adept at using the latest technology, in this book we will not ignore the past. We will learn about transits, theodolites, steel tapes, rods, chains, links, magnetic compasses and other equipment and methods that were used by past surveyors. It has been said that when a surveyor needs to reestablish boundaries that were created a hundred or more years ago, the new surveyor must "follow in the footsteps of the original surveyor". We cannot hope to follow in these footsteps unless we understand the methods and equipment used by the original surveyor.

Even if the reader never needs to interpret a deed written or a plan drafted in the 1800s, having an appreciation of the history will help place modern practices in context. Surveying technology was much simpler when transits rather than GPS receivers were used for making measurements. Surveyors using chains and steel tapes actually made measurements using their own hands. Today, using electronics, measurements are much more abstract because the surveyor is not actually doing the measuring—electromagnetic waves and microprocessors now do the work. For readers unfamiliar with surveying measurements, the relative simplicity of early measurement techniques will provide a solid foundation from which current techniques may be more readily understood.

In some ways, the technical aspects of boundary surveying have gotten easier with the advent of electronic instruments. When compared to the skill required to use a steel tape and theodolite or transit, which took years of experience to perfect, modern instruments and global positioning have made accurate measurements available to nearly anyone who understands the technology. These new capabilities threaten to oversimplify the real challenges faced by today's professional boundary surveyors. Technology has not reduced the required level of knowledge and experience necessary to correctly establish boundaries. Today, a total station or GPS receiver makes it much easier to perform measurements with accuracies far exceeding what was possible using older technologies. But technology does not make it any more likely that a surveyor will locate a boundary at its true location. This is because the legal issues inherent in boundary establishment continue to be the real challenges facing today's boundary surveyors and technology is no substitute for knowledge and experience.

Global positioning has fundamentally changed how boundary surveyors approach their work and many surveyors who formerly relied on total stations for measurements must now be proficient in combining GPS with traditional measurement techniques. This is particularly important when it becomes necessary or desirable to connect a survey to a state plane coordinate system (a way of describing property corners at a precise location in a particular state). However, when the two methods are used together, measured distances are likely to disagree. Surveyors must understand why this happens and how to deal with the differences. Low distortion projections (LDP) are becoming a commonly used way to circumvent some of the limitations and difficulties inherent in state plane coordinate systems, particularly in mountainous areas with high elevations where there can be substantial differences between distances measured on the state plane system and distances measured on the ground.

Some of the chapters in this book contain mathematics, but the concepts are explained in simple terms so that anyone, with a background in high school math, should have no trouble following the logic. The math requirement was intentionally kept to a minimum so that the book would be readable, and hopefully enjoyable, by the widest possible audience. For those who desire a more in-depth treatment of the mathematical aspects of boundary surveying, the Appendix includes the underlying theory and many examples of typical calculations performed by boundary surveyors. The concerns many of us are likely to have about boundary establishment do not require an in-depth understanding of the mathematics underlying boundary surveying. Much of what we need to know will be readily understood if we have a working knowledge of some of the basic technical aspects of surveying and the legal principles of boundary establishment.

Many people need a boundary survey because their boundary locations are unknown or disputed. Even if a property owner believes that he or she knows where their boundaries are located, there are factors such as adverse possession that can change their property lines from the locations stated in their deed. The location and rights associated with easements, such as an old right of way, may be in question. Access easements can have a substantial effect on the value of land. Land isn't worth much if you can't get to it.

Here are some typical questions asked about boundary surveying:

While I was at work, my neighbor had their property surveyed and the surveyor set stakes on my side of the fence. The fence has been there 25 years. My neighbor called me and said they are going to remove my fence. What should I do? Does the neighbor have a right to remove my fence?

My neighbor is having her property surveyed and the surveyor told me that he needed to come on to my property to make some measurements. It's my property and I don't want the surveyor on it. Can the surveyor come onto my property even though I expressly told him to stay off my land?

I found some old stone posts near my lot corners. How do I know if these posts actually fix the location of my property boundaries?

My property does not have any corner markers. I called a surveyor to ask what it would cost to set my lot corners. I think the quoted price is very high. Why does it cost so much just to have a few wooden stakes put in the ground?

A new subdivision has been approved abutting my property, and the bearings and distances shown on the subdivision plan do not agree with those on my deed. Why is the information different? Will recording the subdivision plan affect the location of my boundaries?

There is an old cart path running over my property and my abutter wants to use it for access so he can subdivide his property into 10 house lots. He can't do that, can he?

My lot has concrete markers on each of the corners set by a surveyor when we purchased the property ten years ago. My neighbor just had her property surveyed, and her surveyor put markers two feet from my markers on to the neighbor's property. Now there is a two foot strip of land between my property and my neighbor's property. Who owns this land, my neighbor or me?

Most of these questions do not have simple answers. Some of the answers may be affected by laws in the state where the property is located. The author hopes the following chapters will help the reader to understand some of the underlying issues and how these questions might be answered.

Chapter 2
Getting Started

2.1 Land Surveying Types

A broad definition of land surveying is that surveyors make measurements on or near the surface of the earth. These measurements have traditionally consisted of angles and distances made using optical surveying instruments. Global Navigation Satellite Systems (GNSS or GPS), which use signals emitted by satellites to fix locations on the earth, are supplementing, or in some cases even replacing, traditional techniques and equipment.

Land Surveying can be grouped into several general classifications, such as boundary surveying, topographic surveying, construction surveying, and control surveying. Topographic surveys deal with physical features found at a site and the elevation of the earth's surface. A topographic plan will commonly show buildings, fences, shorelines, sewers and utilities. Topographic plans show contour lines which depict the ground elevations at the site (Fig. 2.1). These plans are often used by developers, engineers and architects for the design and siting of roads, buildings and other structures.

Construction surveys establish control and layout for road and highway layouts, bridges, buildings, sewers and other man-made improvements. Hydrographic surveys are made to acquire data on water depths, locations of rocks, sandbars, coastline erosion and other features associated with oceans, bays, harbors, lakes or rivers.

Control surveying is used to establish precise networks of control points for fixing horizontal and vertical positions. After the control points are established, they can be used for other types of survey projects such as topographic, boundary, route layout and mapping surveys. Control surveys were originally performed using

P. Gay, *Practical Boundary Surveying*, DOI 10.1007/978-3-319-07158-9_2

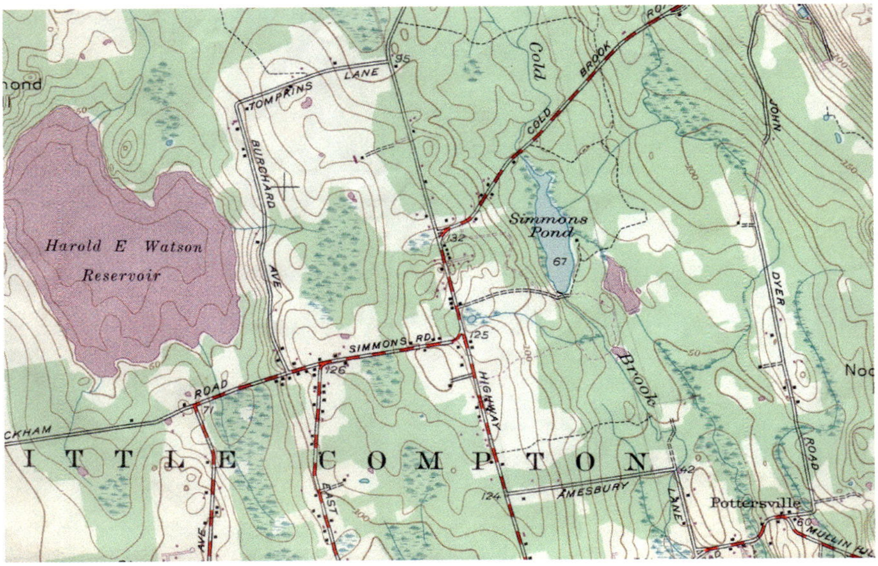

Fig. 2.1 Topographic map

conventional surveying techniques and instruments but modern techniques rely heavily on global positioning because of its high accuracy over long distances.

Cadastral surveying is the determination of the legal boundaries of a parcel of land. It is sometimes called Boundary Surveying or even simply Land Surveying. It encompasses original and retracement surveys of both private and public lands. It is the branch of surveying which will be discussed in this book.

> Cadastral surveying is the determination of the legal boundaries of a parcel of land.

Boundary surveys often include a plan, prepared by the surveyor, showing the results of the survey. A properly prepared plan will show the boundaries of the parcel, and the evidence used to locate those boundaries. The plan will also show the type and location of monuments that were set to fix the boundaries on the ground, and the location of buildings, natural features and improvements on or adjacent to the property. The plan should also show any encroachments and evidence of occupation. The plan is essentially a record of the survey.

Boundary surveys have the potential to affect the legal rights of the property owner and the abutters. Because of this, land surveyors have an obligation not only to their client, but to the public to insure that their survey work is performed with adequate precision and accuracy and that the legal rules governing boundary location have been carefully considered and properly applied.

Land surveying is an ancient profession dating back thousands of years at least to the Egyptians and the great pyramids. The importance of land surveying is not a recent development. Three of the four presidents carved into Mount Rushmore were land surveyors: George Washington, Thomas Jefferson and Abraham Lincoln. History has shown that these presidents were among the greatest presidents to serve our country.

2.2 Land Surveyors Must Be Licensed

Land Surveyors must be licensed by the state in which they practice before they can perform surveying activities. In many jurisdictions, Professional Engineers, such as Civil Engineers, may not establish property lines unless they are also licensed as Land Surveyors.

Land Surveyors must be licensed by the state in which they practice before they can perform surveying activities.

Many states will grant a surveyor who is not licensed in the state but is licensed in another state, permission on a limited basis to perform surveying activities in that state. For example, a surveyor licensed in Rhode Island but not Massachusetts can apply to the Massachusetts Board of Registration of Land Surveyors and Professional Engineers for permission to perform a survey in Massachusetts. If permission is given, it may be limited to performing specific work or work under specific conditions.

States have web sites that list the surveyors that are licensed in the state. If you are in need of a surveyor, you may want to check the state web site to ensure that the surveyor under consideration is properly licensed and is on active status as a land surveyor.

Sometimes a licensed surveyor will send a crew into the field to perform work. On other occasions, a licensed surveyor may have others participate in calculations or research. Under most licensing statutes, the licensed person may delegate some of the work to employees providing that he or she remains in Responsible Charge. Responsible charge means that the unlicensed person is under the direct control and personal supervision of the licensed person. The licensed person is required to supervise the work, review all final documents and take responsibility for the work. One test of whether responsible charge exists is that the client would be able to reasonably presume that the licensed person was the provider of the services. In many if not most cases, the licensed surveyor will sign and stamp plans and other documents with his seal, thereby attesting to the information shown on the documents. Because of the substantial value of real property, licensed surveyors must be very careful not to allow errors or omissions to occur which may have financial consequences for the client, and possibly the surveyor or the survey firm.

2.3 A Land Surveyor's Ethical Obligations

A surveyor has certain obligations to his client and to the public. It is a surveyor's duty to locate boundaries at their true location. Even though a surveyor is paid by his client he has a duty to the **Adjoiner** (abutting property owner) to locate the boundary line in its true location and not in a location that will favor his client's interests. That is not to say that a boundary surveyor shouldn't listen to what a client has to say about boundary locations. For example, a client may recall the exact location of a monument that was destroyed many years ago. The final decision about where to locate the boundary must be based on the legal principles of boundary establishment in the jurisdiction in which the property is located. If the property is in a state in which the public land survey system exists, the rules governing corner reestablishment must be understood and followed.

When re-establishing property lines from a previous survey, a surveyor has a duty to find the original monuments. A surveyor has a duty to inform the client of encroachments on the client's property that come to the attention of the surveyor during the process of surveying the property. A surveyor also has a duty to set sufficient permanent markers so that the client and future owners of the property will be able to locate the property lines.

2.4 What Does a Surveyor Need to Know?

Boundary Surveying requires knowledge of the mathematics and physics necessary to make accurate measurements and to analyze and adjust survey data. It also requires an intimate knowledge of the recording system where property records and road layouts are kept. It requires knowledge of the legal aspects of evidence and procedures for boundary location. It requires diligence in searching for and finding all of the record and real evidence which has the potential to affect ownership and the location of boundaries. It requires an understanding of the surveyor's ethical responsibility and duty to the client and the public. When working in the Public Land Survey System the surveyor must understand the appropriate rules for boundary location and reestablishment as set out by the Government in the *Manual of Instructions for the Survey of the Public Lands of the United States*. The surveyor must be familiar with and understand statutes and case law which have the potential to affect the outcome of a survey. For some surveys, the surveyor must be acquainted with subdivision control laws and zoning.

It cannot be stressed enough that, in many cases, precise measurements, by themselves, will not establish a property line in the correct location if the legal rules concerning evidence and procedures for boundary location are not understood and followed. This is the element that sets boundary surveying apart from a merely technical discipline. This is what makes it a substantially more complex, interesting and challenging endeavor.

It cannot be stressed enough that, in many cases, precise measurements, by themselves, will not establish a property line in the correct location if the legal rules concerning evidence and procedures for boundary location are not understood and followed.

2.5 What a Surveyor Can and Cannot Do

A surveyor can locate a written description on the ground. In some cases the client may provide the surveyor with a written description of the property. In other cases the surveyor will need to search the public recording system for the deed. The property description is found in the deed that the client received when the property was purchased. In most cases the deed will be recorded in the public recording system at the local, county or state level. The boundaries described in a deed are often called **Record Boundaries**. Where property has been inherited, the client may not have a deed in their name and, in such cases, probate records may need to be searched. The deed may make reference to a plan. If so the surveyor will consider both the plan and the deed as providing a description of the property to be surveyed.

Although it was stated earlier that surveyors establish property lines, surveyors cannot always determine the actual extent of ownership with finality. In some situations the client's deed description accurately describes the extent of ownership, but in many cases it does not. A property owner can always claim to own all of the property described in their deed. However, there are many factors that can modify the record boundaries of ownership. For example, unwritten title, such as adverse possession can change the location of record boundaries. Prior conveyances by a common grantor can change the record boundaries. Mistakes in the survey that created a series of subdivision lots can effectively change the record boundaries of the subdivision lots.

When it becomes apparent to a surveyor that the record boundary locations conflict with other evidence of ownership, or there are conflicting deed elements, all the surveyor can do is to establish the boundaries on the ground in accordance with the legal rules governing boundary establishment and advise the client to seek advice from an attorney on how to proceed in order to resolve the issues. It is important to recognize that boundary locations are based on the surveyor's opinion of the available evidence and application of the appropriate legal rules. In such cases, unless the parties can come to an agreement over the boundary location, litigation may be the only solution. Each client must recognize that, even though they may have paid their surveyor a substantial fee for locating a boundary, the location is really just an opinion. It does not have the force of law. This will be discussed in more detail when we look at the surveyor's authority.

2.6 The Surveyor's Authority

Many original boundary surveys are imperfect in one sense or another. Original monuments become disturbed over time, errors may be discovered in measurements, plans or deed descriptions. Perhaps the original surveyor was unable to find all the evidence, made incorrect decisions or interpreted evidence incorrectly. Sometimes these errors or problems are small and will have very little practical effect on the survey. It is not uncommon, however, for the effects to be substantial and the results unacceptable under the circumstances. Because even the most precise surveying techniques always involve some degree of error it can be safely said that all retracement surveys (resurveys of property originally surveyed) will result in some disagreement between the original survey and the new survey.

Theoretically, if all surveyors were successful in discovering all the evidence, had equipment of similar precision, made measurements and adjustments using proper techniques and correctly applied the legal rules and procedures for boundary location, there should be little or no disagreement between the resulting boundary locations. In the real world this may not actually happen, and three surveyors may come up with three different locations for the same property corner. With the precision and accuracy available with modern measuring equipment, most surveyors shouldn't have much disagreement over their measurements. Interpreting the evidence for boundary establishment, however, can result in substantially different opinions, which can lead to large differences in boundary locations.

There is no law requiring each surveyor to arrive at identical boundary locations. Each surveyor is entitled to his or her opinion regarding where they believe the true boundary should be located. To be sure, theirs is an educated opinion and one based on experience, but it is an opinion nonetheless. It might be argued that a surveyor working in the public land survey system will have an easier time with troublesome boundaries than a surveyor in the eastern colonial states because the rules for boundary relocation have been more clearly stated by the government. However, surveys of the public lands present their own unique set of challenges.

Unless the surveyor is working under the authority of the judicial system and is performing a survey for or under the direction of a court of competent jurisdiction, the boundaries established by the surveyor do not have the force of law and are subject to relocation. In most circumstances, the only way to fix boundaries permanently is by court order or by agreement between abutters. So, while a surveyor can locate a boundary on the ground, the location is really the surveyor's opinion of the true boundary location.

Unless the surveyor is working under the authority of the judicial system and is performing a survey for or under the direction of a court of competent jurisdiction, the boundaries established by the surveyor do not have the force of law and are subject to relocation.

2.7 Trespass by Surveyors and Adjoining Property Owners

Can a surveyor go onto abutting property without permission of the land owner? **Trepass** is the intentional intrusion or invasion of the land of another person. Any knowing entry is a trespass. It has been said that the essence of private property is the right to exclude others. Because private property is so highly regarded in our legal system the courts have been reluctant to diminish the right of an individual to keep unwanted persons from entering their property. Liability is incurred by the trespasser merely by knowingly entering the land of another. It is not necessary that the trespasser actually damage the owner's property. Just being there is enough (Fig. 2.2).

A person can commit a trespass not only by their presence on the land of another, but also by causing some object to enter upon the land of another. Setting a concrete bound on the abutting property without permission of the landowner is a trespass. Erecting a building or putting a driveway on abutting property, without permission, is a trespass. When a surveyor cuts brush on an abutter's property without consent of the owner, the surveyor can be held liable for trespass and for damaging the owner's property. A machete which is commonly used by surveyors for cutting brush to clear survey lines can be seen in Fig. 2.3.

Trespass can also be a criminal offense. This commonly occurs when the trespasser knows they are not allowed on the property or when the trespasser remains on the property when told to leave. Entering upon posted property or upon property after being told by the property owner not to trespass may subject

Fig. 2.2 Surveyors must understand the laws against trespass prior to entering upon abutting property

Fig. 2.3 A machete is used for cutting brush when surveying in the woods. Surveyors must be careful not to cut on abutting property without consent of the owner

the trespasser to criminal charges. Criminal trespass is considered a crime against the community and as such can be punishable by fine or imprisonment or both.

Some jurisdictions have enacted statutes allowing surveyors to enter upon lands adjoining the land being surveyed in order to conduct surveying activities. Usually the surveyor is required to give reasonable notice to the adjoining land owner. Once proper notice is given the entry becomes privileged and the surveyor may not be held liable for trespass. Unless a jurisdiction has such a statute a surveyor has no more privilege than anyone else; the surveyor cannot enter upon land without permission of the owner and doing so will subject the surveyor to liability for trespass. Even where there is privilege to enter upon adjoining lands the surveyor has no right to cut brush or otherwise damage the land owner's property.

Some jurisdictions have enacted statutes allowing surveyors to enter upon lands adjoining the land being surveyed in order to conduct surveying activities.

2.8 Why Become a Land Surveyor?

Boundary surveying is a great profession, and most surveyors remain in their profession because they love their work. There is a great deal of variety in the work because every boundary survey is different. Boundary surveyors are always visiting new places. They are constantly meeting new people and solving new problems. When the surveyor is not doing research, he or she may be performing field work. When the field work is done, the surveyor may be in the office performing calculations or meeting with clients. Surveyors often attend public meetings in order to have their plans approved. They will often meet with local building inspectors, engineering departments, health boards and conservation agencies regarding surveys and plans.

Boundary surveyors must be adept at making very precise measurements in difficult environments and under difficult working conditions. Surveyors spend a lot of their time in the field. Much of the work is physically demanding so surveyors must remain physically fit. Setting lot corners on their client's property can be hard work. Setting a concrete or stone marker means digging holes three feet deep. If these markers are far in the woods, the survey crew must carry in the heavy markers and the equipment used to set them. There is also equipment to maintain. Working outdoors presents many challenges. Surveyors are sometimes working in the woods where they have the opportunity to observe wildlife and be observed by wildlife. Sometimes in remote areas, particularly where there are bears, mountain lions or poisonous snakes, wildlife encounters can be potentially dangerous and surveyors must take appropriate precautions. In some remote areas, where wildlife is known or likely to present a danger, it is prudent for the surveyor and the crew to carry firearms or pepper spray. There are also lots of bees, ticks, mosquitos and gnats to contend with. When working on farms and ranches, there are usually domestic animals. Some are naturally curious and insistent on visiting the survey crew—whether they are welcome or not.

Boundary surveyors must be adept at making very precise measurements in difficult environments and under difficult working conditions.

Getting equipment in and out of remote locations can require the use of all-terrain-vehicles (ATV), snowmobiles or horses. An ATV set up to carry surveying equipment can be seen in Fig. 2.4. In very remote places, equipment and people must be flown in by aircraft or float plane. Measuring across streams is easy with electronic measurement equipment but the survey crew will still need to cross the streams to bring equipment across. If the water is deep, a boat may be the only way to cross. Sometimes the boundary marker that has to be located is in the middle of a swamp and one or more of the crew has to cut line through the brush to get to it. There is often poison ivy or poison oak, so surveyors need to be skillful in plant species identification. In inaccessible mountainous terrain packing equipment in by foot or horseback may be the only practical and economical alternative.

Fig. 2.4 All terrain vehicle (four-wheeler) set up to carry surveying equipment in remote areas or in large acreages. Notice the instrument tripods mounted in a special fixture

Many surveys are on a budget, so hiring a helicopter at several hundred dollars per hour might be out of the question.

Surveying in urban and suburban locations also presents challenges. In these areas, most properties are bounded by streets and highways so it is necessary to run survey lines in places with vehicular traffic—often fast moving traffic. Surveyors working in these locations soon come to realize that the inattention of some drivers poses a substantial hazard to those working on or near a traveled roadway. Even with proper precautions such as traffic cones and signs, constant vigilance is required. Working on the relatively benign area of a sidewalk can also subject a surveyor to hazards. Ask the surveyor whose ten thousand dollar total station was tipped over onto the concrete when the training wheel of a little boy's bicycle happened to catch the tripod leg.

The weather can also be challenging. Working in hot sunny climates, surveyors have to deal with heat shimmer which refracts light and makes accurate sighting of points through the instrument a real challenge, particularly when sighting over long stretches of pavement. In cold climates, snow covers physical evidence making boundary markers challenging to find. Frozen ground makes it difficult or impossible to dig in order to find buried points. Snowmobiles, skis or snowshoes may be required in order to get from one point to another. Rainy climates pose their own challenges in keeping equipment dry and serviceable, not to mention keeping the crew dry. Sometimes, the survey crew doesn't have a great deal of choice other than to accept the weather and get on with the work.

There are also hostile abutters. Some of the work performed by boundary surveyors is the result of disputes over one or more boundary lines. By the time a surveyor is retained, relations between the parties may have deteriorated to the level where the only way of performing the survey is by having the local police on hand to keep the peace. Because a surveyor is working for an adversarial party, abutters sometimes view the surveyor in a less than favorable light. It is not always well understood that the surveyor's job is to locate a boundary in its true location. A surveyor cannot favor a client's interests; he must remain neutral. Nevertheless, the fact that the surveyor is "working for" the adverse party is often enough to tip the scales against him. This can make the performance of the survey difficult, particularly if the surveyor needs to go onto the abutting property to look for or measure to physical evidence. In many cases, the surveyor's application of a bit of "finessing" can ease what would otherwise be an awkward or even dangerous situation. In addition to understanding how to perform a survey, a surveyor sometimes needs to understand and employ a bit of psychology.

Because surveyors work in public places, they are often asked questions by passers-by. Some of the most common questions to the author are: "Are they going to fix the road?" (This may be more of a wish than a question.). "What are you taking pictures of?" (Total stations and theodolites must look like cameras to some people). "I live in that house down the street. What would it cost to put my markers in?" (This is often impossible for the surveyor to answer without first doing some research). The author recalls one survey where an abutter walked over and asked "What are you doing?" When I answered that we were surveying her neighbor's land, she responded that in her experience there were two kinds of people that she feared most: "surveyors and lawyers". I didn't have the heart to tell her I was both.

Every survey presents new challenges. Even after years of surveying, when you may think that you have experienced every situation, a new challenge will present itself. The reader may ask, with all of these potential difficulties and hazards, why would anyone want to become a boundary surveyor? The best answer is: That is precisely what makes the work so interesting and worth doing.

Chapter 3
Basic Legal Concepts

Before we can examine the process of performing a survey we need to have an understanding of some concepts which are fundamental to boundary surveying. Boundary surveying involves both legal and technical concepts. In the next couple of chapters we will discuss evidence, the definition of a boundary, distances, angles, bearing systems, the concept of property line closures and the technology that enables surveyors to make precise measurements.

The establishment and reestablishment of boundary lines requires the surveyor to have a working knowledge of certain aspects of the law of evidence, so we will begin there.

3.1 Evidence

Evidence is legally defined as proof legally presented at the trial of an issue, by the parties, in the form of witnesses, records, documents, exhibits, objects, etc., for the purpose of ascertaining the truth or untruth of some matter under consideration. In this book we will remove the requirement of a legal proceeding from the definition and use the word evidence in a broader sense. Evidence is something that furnishes proof for the purpose of ascertaining the truth or untruth of some matter or claim. Evidence can take the form of statements by witnesses, records, documents or objects if these furnish proof.

> Evidence is something that furnishes proof for the purpose of ascertaining the truth or untruth of some matter or claim.

Although most boundary surveys do not end up in court, the importance of evidence should not be diminished by that. All surveyors should approach each survey

P. Gay, *Practical Boundary Surveying*, DOI 10.1007/978-3-319-07158-9_3

as though it were under the scrutiny of an adverse party. It is the surveyor's duty to exercise diligence in discovering all evidence applicable to the survey. Clients should be informed that, in resurveys of existing property, it is not always possible, at the beginning of a survey, for a surveyor to know how much time will need to be expended in order to recover the evidence necessary to competently perform the survey. Evidence called for in old surveys is often missing, and reestablishing the location of old markers is often very difficult and time consuming. In many surveys, finding evidence is by far the most difficult challenge facing the surveyor. Actually setting the lot corners can be the quickest and easiest part of the work.

Surveyors need to know about three types of evidence: **Parole Evidence (oral evidence)**, **Written Evidence** and **Physical Evidence**. Parole evidence is testimony given by witnesses such as a surveyor, property owner, abuttor or someone knowing something about the property at issue—it is spoken evidence. Parole evidence is also called oral evidence. Written evidence consists of documents such as deeds and plans. Physical evidence, sometimes called **real evidence**, consists of physical objects such as concrete bounds, drill holes, stone walls and waterlines.

3.2 Record and Physical Evidence

A surveyor most often works with two basic types of evidence, **Record Evidence** and **Physical Evidence**. Record evidence is a form of written evidence. In its broadest sense it consists of documents recorded or held by a public agency charged with the responsibility of maintaining such records. Deeds and plans recorded in a recorder's office, office of the city clerk, or sometimes the town highway department, are forms of record evidence.

Physical evidence is used in deeds as a means of locating property corners on the ground. For example, a deed may say that a property line begins at a stone bound then runs along the edge of a stream. Both the stone bound and the stream are physical evidence. A deed describing property would not be very useful if it did not contain information that allowed the property owner to locate their boundaries on the ground. In fact some deeds do not contain references to physical evidence. In such cases it can be very difficult to establish the property lines with a reasonable degree of certainty.

3.3 Implied Notice of Recorded Documents

When deeds and plans are recorded in a recorder's office the public is presumed to be on notice of the contents of the documents, even if a person has not actually seen or read them. For example, suppose you are purchasing a parcel of land. The seller presents you with a deed describing the land. Before you purchase the property you will want to go to the recorder's office and do a title search to ascertain if the seller has already sold all or some portion of the property to someone else. You also want to know if there are any liens or encumbrances on the property. If you purchase the property without doing a title search and later find out that a portion of the property was sold several years ago to another party, and that party recorded a deed, you may not have a valid claim to that portion of the property previously conveyed.

3.4 Property Lines and Boundaries

A Property Line or Boundary is the limit of ownership of land held by title deed or by marks of enclosure if possession is without title deed. A title deed is merely the deed recorded in the recorder's office that describes the property in question. Where land is held by title deed, boundaries may be described in the deed by measurements such as distances, angles and bearings and by reference to monuments.

It is also possible to own property or to have rights in the property of others even though there is no writing to evidence the ownership. Such title is often referred to as **Unwritten Title**. Adverse possession is a species of unwritten title. Adverse possession is acquiring property by occupying the property of others in accordance with certain statutory or common law requirements. Although title to property is acquired there is no deed to evidence the transfer. Unwritten title will be discussed in a later chapter but for now, the reader will understand that a writing does not always accurately describe the actual extent of ownership.

A Property Line or Boundary is the limit of ownership of land held by title deed or by marks of enclosure if possession is without title deed.

3.5 Physical Monuments

Most deeds call for monuments in one form or another. There are three commonly recognized types of monuments: Natural Monuments, Artificial Monuments and Record Monuments. Natural and artificial monuments refer to physical things whereas a record monument is more abstract. Natural monuments are relatively permanent things naturally existing in or upon the ground (placed there by nature) such as the edge of a river, lake or pond, trees, and boulders. Artificial monuments are things constructed or placed in or upon the ground by man. Fences, stone walls, piles of stones, stone and concrete bounds, wooden stakes, drill holes and iron pipes are all artificial monuments (Figs. 3.1, 3.2). Some artificial monuments such as concrete or stone bounds are considered permanent while others, such as

Fig. 3.1 Copper pin—artificial monument

wooden stakes and pipes are not because they may deteriorate over a short period of time or be easily moved.

There are three commonly recognized types of monuments: Natural Monuments, Artificial Monuments and Record Monuments.

3.6 Record Monuments

Record monuments, sometimes called legal monuments, are monuments that are called for in a legal description recorded in a public recorder's office, for example in a registry of deeds or in the office of a town clerk. If your deed reads that your land is bounded on the north by land belonging to "Samuel Smith", then Mr.

Fig. 3.2 An old stone bound—an example of real, or physical evidence

Smith's property line would be a record monument. If your deed calls for your east line to run along Main Street, then the line of Main Street would be a record monument.

Record monuments are potentially more difficult to establish on the ground than are physical monuments. If you deed calls for a stone wall as your southerly boundary and if the wall is still there and it is the same wall that was in existence when the original deed was created, it should not be too difficult to establish your property line. The line is wherever the wall is located.

But what if your deed just says that your north line is bounded by land of Samuel Smith, making no reference to physical monuments? How do you locate your boundary? Your deed is actually saying that in order to establish your property line you must establish Samuel Smith's property line. Your next step would be to go to the recorder's office and read Samuel Smith's deed. What if Samuel Smith's deed description makes no reference to physical monuments either, it just calls for you as the abutter? The lack of a call for physical monuments in the deeds will probably make it difficult to accurately establish the boundary on the ground.

Let's go back to the portion of the hypothetical deed that calls for your east line to be the line of Main Street. Research discloses that Main Street is a public street accepted by the city long before your lot was created. After obtaining layout plans for Main Street from the recorder's office you find that the plan calls for stone monuments along the side lines of Main Street. These stone monuments become controlling in establishing your east property line. The record monument called for in the deed was the line of Main Street. The line of Main Street was established by the city by acceptance of a public layout. The layout plan calls for stone monuments. The call in your deed for Main Street is effectively a call for stone monuments. If the original monuments are still in existence, their location will control the location of your property line along the street.

Chapter 4
Basic Technical Concepts

4.1 Distance and Direction

The accurate description of boundary lines requires that each boundary be described as having a certain finite length, and a known direction. In a deed, the description of boundaries is referred to as the **Metes and Bounds**. Metes are the courses and distances of the boundary lines. Bounds are the natural or record monuments which fix the lines.

The direction of a boundary line may be fixed using **Angles**, **Azimuths**, or **Bearings**. Bearings are the most commonly used form of direction. Angles are sometimes used. Azimuths are less commonly used. In some cases, no distance or direction is given, and boundaries are simply described by reference to monuments at the beginning and end of one or more lines.

4.2 Distances

In boundary surveying, distances are always Horizontal Measurements, usually measured in feet. In very old surveys, distances are sometimes stated as Rods, Chains or Links. In the Public Land Survey System, the official units of measure are chains and links. In the metric system, the standard length used in surveying is the meter.

> In boundary surveying, distances are always Horizontal Measurements

In boundary surveying, division of the foot (or meter) is made using the decimal system not the fractional system. For example, a distance of 100.50 feet would be equal to one hundred feet, 6 inches. Inches are almost never used by surveyors. If you run across a deed having dimensions described in inches you would probably

P. Gay, *Practical Boundary Surveying*, DOI 10.1007/978-3-319-07158-9_4

be safe in assuming that the measurements were not made by a land surveyor. Most modern survey plans show boundary dimensions to the nearest hundredth of a foot. For example 100.75 feet contains two decimal places. The first place right of the decimal point indicates tenths of a foot and the second place hundredths of a foot. A hundredth of a foot is about 1/8 of an inch (1/8″ = 0.0104′). For measurements to objects where the surveyor wants to indicate less precision, dimensions can be rounded to the nearest tenth of a foot (0.1) or even to the nearest foot. The precision of a number, when noted on a plan for example, tells the reader something about the degree of confidence that a surveyor has in the measurement. For example, if the surveyor is measuring to the center of a roughly constructed field stone wall which might be 3 or 4 feet wide, showing a dimension to the nearest hundredth of a foot may be overstating the surveyor's ability to determine the exact center of the wall. A dimension having a precision of 0.01 will indicate a much higher level of confidence than one shown to the nearest foot. This notion will be clarified in the section which discusses precision and accuracy.

When surveying calculations are performed, it is desirable to use additional decimal places in order to increase precision and minimize rounding errors. The use of 3 or 4 decimal places is common for distances and an even larger number when using trigonometric functions. Electronic calculators and software operate internally using a large number of decimal places, even though only two or three places may be displayed.

All distances in deeds and in surveying are assumed to be **Horizontal Distances** unless noted otherwise. The term horizontal means a level line, a line parallel to the horizon. When measuring distances with a steel tape, particularly when measuring up or down a hill, a level must be used to verify that each end of the tape is held at the same elevation. A horizontal measurement is illustrated in Fig. 4.1. When there is a clear view of the horizon, for example along the ocean shore, it is possible to sight along the tape and raise or lower the ends until the tape is in line with the horizon. The tape will then be sufficiently level to make an accurate horizontal measurement.

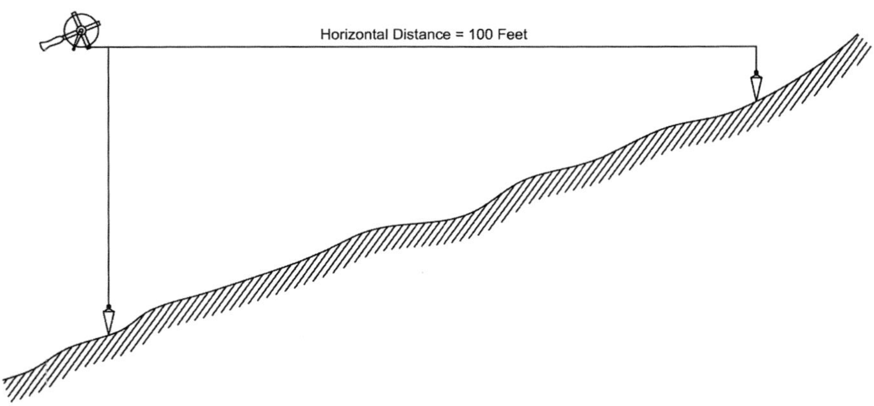

Fig. 4.1 Horizontal measurement

Sometimes on a very steep hill it is not possible to hold the tape level. Even if the uphill person holds the end of the tape on the ground the hill may be so steep that the downhill person may not be tall enough to raise the far end of the tape to a level position. In such cases the tape can be held at some convenient angle, perhaps parallel to the slope of the hill. The angle of the tape from horizontal must be accurately measured and recorded along with the distance. The angle is called a **Vertical Angle**, because it is measured in the vertical plane and the distance is called a **Slope Distance** to differentiate it from a horizontal distance. Once the vertical angle and slope distance are known, the horizontal distance can be accurately calculated. It can be seen from Fig. 4.2 that the horizontal distance is always shorter than the slope distance.

When measuring up or down a steep hill, if it is not possible or convenient to measure the vertical angle of the tape, it will be necessary to **Break Tape**. Breaking tape means that instead of using the full length of the tape, 100 feet for example, a shorter distance is measured. In the previous example, 50 feet could be measured twice, as in Fig. 4.3. Using a shorter tape length decreases the vertical distance making it easier to complete the measurement.

Surveyors carry a handheld sight level when taping to ensure that the tape is held level. In an emergency, if the level were to become lost or broken for example, it is possible to tell if the tape is level by the measurement itself. For example, when measuring between fixed points, such as two stone bounds, a series of measurements can be taken with the one end of the tape held at varying heights. When the tape gives the shortest distance between the fixed points the tape will be level and the distance can be recorded. When setting out a point just the opposite procedure is used. One end of the tape is raised and lowered until the distance between the points is the greatest.

Most steel tapes in common use for land surveying in the U.S. are 100 feet in length. In order to make an accurate measurement the tape must be pulled tight with a certain amount of tension. The tension reduces the amount of sag in the tape. The amount of tension depends on the particular tape, but for a 100 foot tape, the tension

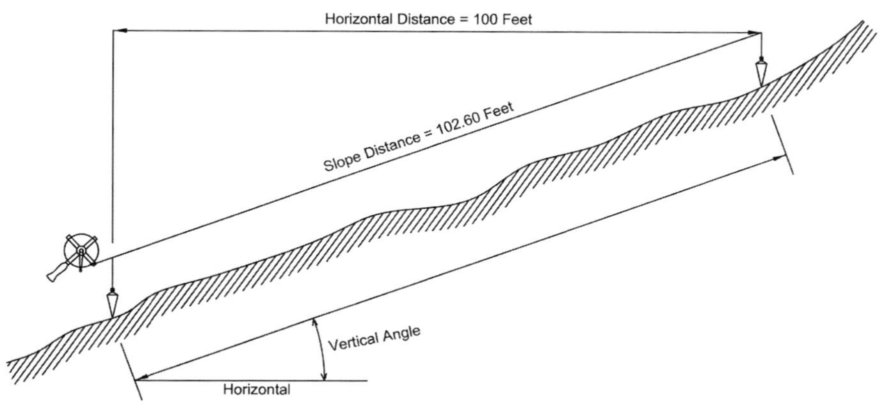

Fig. 4.2 Measuring a slope distance with a tape

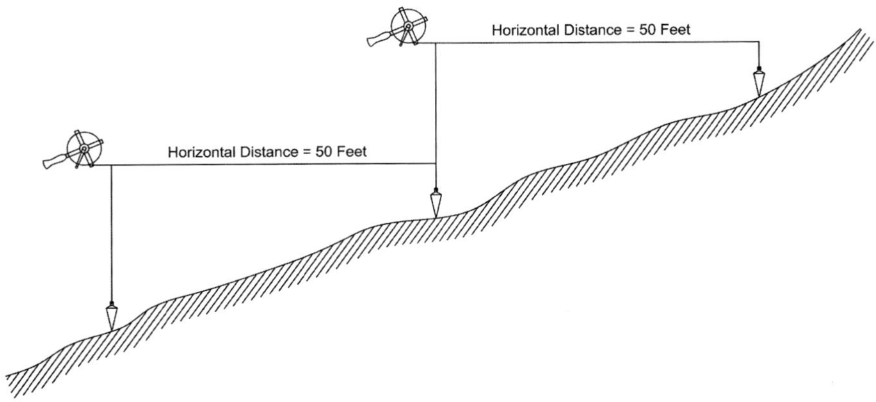

Fig. 4.3 Breaking tape—measuring short distances on a steep slope

can be about 25 pounds. Rather than guessing at the tension, it is best to use a tension meter (spring balance) on one end of the tape. When measuring up or down a steep slope the person holding the tape at or over their head must be strong and in good physical shape in order to resist the pull on the tape while holding it steady enough to obtain an accurate measurement. The amount of tension required is dependent on the tape length, so long tape lengths require considerably more tension than shorter ones.

When measuring long lines greater than the tape length of 100 ft, it is necessary to set intermediate points on the ground so that there will be a point from which the next 100 ft measurement can be made. For example, if a 200 ft line were being measured, the front person would set a point at exactly 100 ft. The next 100 ft would then be measured from that point. The point could be a nail pressed into the soil or a mark set into or on the pavement. When taping was the prevailing method of measurement, many surveyors used "**Surveyor's Arrows**" or "**Chaining Pins**". These are hardened steel pins a foot or so long, often painted in red and white bands. There are 11 pins in a set. Although measuring survey lines with a steel tape is no longer very common, the pins can still be purchased from surveying supply houses. When a long line is to be measured, the front person sets a pin at exactly 100 ft by pressing it into the soil. Once the front pin is set, the rear person then removes the rear pin from the soil and keeps it. When taping long lines it is easy to forget to count one or more 100 ft increments, so after the complete line is measured the number of pins carried by the rear person is counted. The number of pins multiplied by 100 ft, plus the remaining distance to the end point (a partial tape length) is the total distance of the line being measured. For example if a line of 535.23 ft were measured, there would be five pins and a partial tape length of 35.23 ft. The above assumes that the starting and ending points are existing survey points such as stone bounds or wooden stakes, so no pins are used for these points.

Steel tapes used by surveyors are usually wound on a metal reel as shown in Fig. 4.3, which shows two measurements down a steep slope. The reel has a winding handle which also serves as a lock to prevent the tape from unwinding. When making measurements less than 100 ft, the tape is wound onto the reel so that only the length of tape required to make the measurement is exposed. The end of the tape has a loop and often a leather thong which assists in holding the tape against the tension. When taping 100 ft increments along a line, the tape is unrolled to the 100 ft mark and the handle locked. The front person carries the reel. After the front person sets a chaining pin, the rear person drops the end of the tape to the ground. The front person holding the reel then drags the tape along the ground while walking the next 100 ft. In practice, the rear person should walk slightly faster than the front person in order to arrive at the chaining pin first, so he can announce to the front person that the end of the tape has arrived at the pin (often by shouting the word "Tape"). The rear person should never attempt to carry the end of the tape—it should be allowed to drag over the ground. Steel tapes are quite brittle and if the tape should form a loop and kink, it will break. By allowing the tape to drag over the ground a kink is less likely to occur than when both ends are restrained, unless the tape is constantly held under tension while walking (usually an impractical and inconvenient alternative).

When taping between points that are more than 100 ft apart, the chaining pins must be close to the line or the distance measured along the line will not be correct. For example, if one end of a 100 ft measurement were 1.5 ft off-line, the distance along the actual line would be 99.99 ft, not 100.00 ft. The tape can be kept on-line when measuring by having the instrument person give direction to the front person as each chaining pin is set. This requires a three person crew—the instrument person and two persons taping. If the crew consists of only two persons, an alternative method of taping is to set random points on-line using the instrument and then taping between the points. Of course, the points must be less than 100 ft apart. One way of setting the points is by pacing off distances considerably less than 100 ft. For example, if a person consistently takes 33 paces in order to travel 100 ft, points could be set on-line at 28 or 30 paces, thereby insuring that the points will be less than 100 ft apart.

The disadvantage of using partial tape lengths for all measurements is that the tape must be read each time a measurement is made. Because all of the distances will be fractional tape lengths, a mistake in reading the tape or transposing the numbers when writing them in the field book is much more likely than when working with exact 100 ft increments. For example, a distance of 89.58 ft might be transposed and written as 88.59 ft. Where 100 ft measurements are made, only the last distance will be less than 100 ft. The chance of transposing a number is therefore much less because the number of fractional tape lengths is less. Additionally, when adjusting a traverse, the location of an exact 100 ft error (blunder) is much easier to find than an error caused by reading a tape incorrectly or writing an incorrect foot or decimal portion of a foot. When performing retracement surveys where errors in the original survey are discovered, access to the original field books, in which each taped distance should have been recorded, may provide clues as to the likely source of the error.

Accurate measurements with a steel tape also require **Temperature Adjustments** to the distance measured. A steel tape will expand when hot and contract when cold. Tapes are calibrated at 68 °F (20 °C). For a change in temperature of 15 °F, a steel tape will change in length about 0.01 ft, or about 1/8 of an inch.

Most distances these days are measured using an **Electronic Distance Measuring Device** (**EDM**) attached to a total station rather than a tape. The widespread use of EDMs does not, however, excuse surveyors from knowing how to use a steel tape to make accurate measurements. Electronics sometimes fail, batteries go dead and instruments lose their calibration. In some jurisdictions, for some types of surveys, regulations may require that the surveyor use a steel tape to confirm that electronic measurements are as accurate and precise as the equipment manufacturer declares them to be and that the equipment is properly calibrated. As the above introduction to using a steel tape shows, making accurate measurements with a steel tape requires substantial experience and great care. Notwithstanding the difficulties, during the era when steel tapes were used for most surveys, experienced surveyors routinely made extremely precise measurements.

4.3 Distances: Units of Measure

Old deeds and plans sometimes give measurements in **rods**, **chains** or **links**. In the sixteenth century a rod was defined as the length of the left feet of the first 16 men out of church on a certain Sunday. Sometimes a rod was called a "rood", "perch" or "pole". Today, the standard length of a rod is 16.5 feet. Prior to the twentieth century, property was often measured by dragging a surveyor's chain along the ground. Although it is known that the chains used by various surveyors were of slightly different lengths, the chain is now commonly accepted as being 66 feet, or 4 rods in length. The chain was composed of 100 links, each link being 0.66 feet in length. Except for the public land survey system, chains are no longer used in surveying.

In the sixteenth century a rod was defined as the length of the left feet of the first 16 men out of church on a certain Sunday.

Dragging a chain over the ground will not be likely to give the true horizontal distance between points, particularly when the ground is uneven, when trees are in the way or when measuring on a hill. Some surveyors were known to add a little bit to each chain length, "for good measure." When old deed descriptions, showing distances in rods, are laid out using modern equipment, there are often differences between the record distances and the physical evidence. Surveyors are aware that the unknown length of an original surveyor's chain and the measurement techniques that were used are likely to yield inconsistencies between the original measurements and modern measurements. We will see in later chapters how the law deals with these differences.

In most states, when referring to surveying activities, the **Survey Foot** is used as the standard unit of measurement. In a few states, the **International Foot** is used. A couple of states do not specify which one should be used. The survey foot is defined as:

$$Survey\ Foot = \frac{1{,}200}{3{,}937}\ \text{m}$$

The difference between the survey foot and international foot is small (2 parts per million) and for most small boundary surveys the differences between the survey foot and the international foot will not matter because measurement errors will exceed the difference. However, when working on surveys in which measurements may be miles in length, the differences become more important.

There have been attempts to change surveying measurements in this country to the metric system. The metric system is commonly, almost exclusively, used by the scientific community. During these attempts some boundary surveyors did show both feet and meters on their plans. Showing two sets of dimensions is cumbersome and when you write the something twice you are twice as likely to make a mistake. As it turned out, almost no one was interested in writing deeds with distances measured in meters. Most boundary surveyors reverted to using feet exclusively. The reader will have noticed that the survey foot is actually defined in meters. However, it is still common for deeds to call out dimensions in feet. The tradition of using feet is so well entrenched in the law of property that it will probably require legislation before the metric system becomes widely used in this country for boundary surveying.

4.4 Area

Area is the amount of space contained within a two dimensional geometric figure. In the United States, area is usually measured in square feet or acres. Legend has it that an acre was the amount of land that a yoke of oxen could plough in one day. Today, one acre contains 43,560 square feet of land.

Consider a simple example of calculating area. A perfectly rectangular room measures 10 feet by 12 feet. It would contain 120 square feet ($10 \times 12 = 120$). In this example, area can be thought of as the number of 1 foot square floor tiles that would cover the floor. Note, that if the room were not rectangular, but was skewed into a parallelogram, the area would be less.

Area in boundary surveying is always measured in the horizontal plane. A square parcel of land measuring 100.00 feet by 100.00 feet would contain the same area (10,000 square feet.) on a level and flat meadow as it would on a steep mountain side. Although the actual surface area of a steep mountain side would be substantially greater than that of the flat meadow, in boundary surveying, the area of both are considered to be the same.

If a parcel of land contains 190,000 square feet, it could be described as containing "4.36 acres" or "4 acres and 15,760 square feet". For smaller parcels, containing a few acres or less, acreage is often stated in square feet. For large parcels it is usually described in decimal acres or acres with the remaining fraction in square feet.

In conveyances, area is often described as "more or less". In the previous example, a deed may use language such as "containing 4.36 acres more or less". The "more or less" language has its origins in times before calculators and computers were available for calculating areas. The calculation of the area of an irregularly shaped parcel is cumbersome and time consuming to perform by hand. Because of this difficulty, it was common before computers became widely available, to use a **Planimeter** such as the one shown in Fig. 4.4. A planimeter is a mechanical device that is capable of measuring area. In operation, the planimeter is placed on top of a drawing which depicts the boundaries of the parcel for which the area is to be determined. One of the arms contains a needle pointer or magnifying glass at its end. The pointer is moved around the entire perimeter of the parcel, beginning at a particular point and ending exactly at the starting point. There is a dial which rotates and records the amount of movement. The dial is read at the beginning and end of the tracing movement and the difference in the reading is multiplied by a scale factor to calculate the area.

A Planimeter is a very small, desktop, device with a limited coverage area. It is therefore necessary to reduce the plan being traced to a size small enough so that the planimeter can trace the perimeter of the parcel. The larger the parcel of land, the greater the reduction in size and the less accurate the area determination becomes. It is possible to use a larger plan by breaking it up into a number of smaller pieces and adding the areas together. Nevertheless, measuring areas using

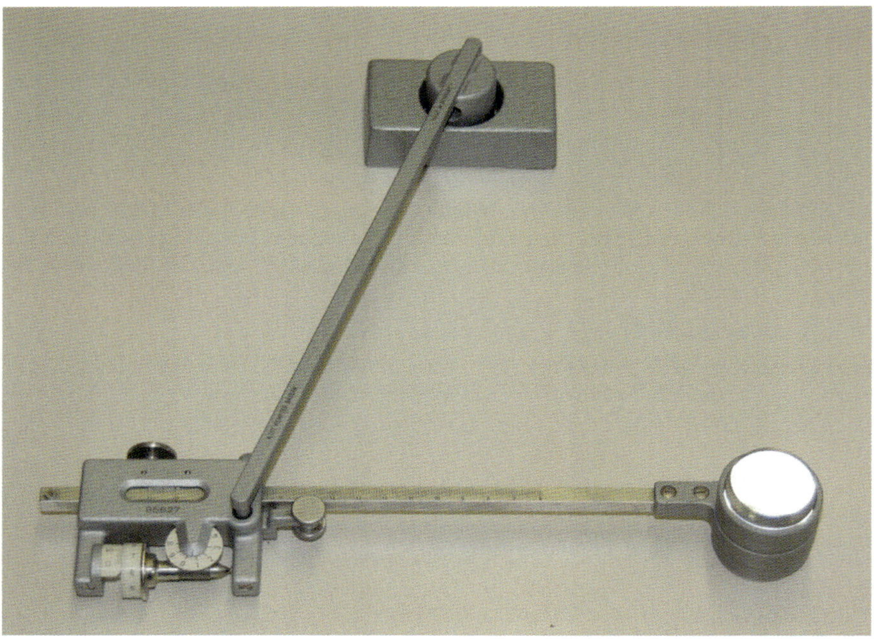

Fig. 4.4 Polar planimeter—creative commons license

a planimeter will not yield the accuracy of areas calculated using the mathematical methods that computers have made possible.

There is case law which suggests that when a person purchases a parcel of land and subsequently learns that the area actually conveyed is less than that called for in the deed, the seller may be liable for damages and reimbursement for the value of the shortage. If the seller uses "more or less" language he or she can claim that the exact area was unknown and the buyer agreed to take whatever actually existed. Such a claim may or may not prevail in court, and indeed, there is case law to suggest that such a claim should not prevail. However, if the actual area was unknown or uncertain at the time, it at least provides the seller with an argument.

These issues have largely been eliminated by the use of computers. It is now possible to calculate the area of an irregularly shaped parcel in surveying software by merely clicking the mouse inside the parcel. If the software is also used for drafting the final plan, it is not even necessary to write down the area as the software will annotate the drawing with the calculated area. The area of a parcel remains an important element in its value so it is important for any surveyor depending on a computer calculated area to still apply common sense and scrutinize the calculated area to determine if it is reasonable. When interpreting old plans and deeds it is important to recognize that the stated area may not be representative of the actual area.

4.5 Angles

An **Angle** is a measurement that describes the geometric relationship between two intersecting lines. In boundary surveying in the United States, the **Sexigesimal System** of angle measurement is used. The sexigesimal system is based on the number 60. Angles are measured in **degrees**, **minutes** and **seconds**. There are 360° (degrees) in a circle, 60′ (minutes) in a degree and 60″ (seconds) in a minute. As can be seen from Fig. 4.5, angles are always measured from a reference line and can be measured to the right (clockwise) or to the left (counterclockwise).

Figure 4.6 shows a parcel of land described by angles and distances. Note that the angles show the relationship of two lines that intersect. These intersections are labeled A–D in the figure. Using the angles and distances, the parcel becomes a rigid two dimensional geometric figure.

4.6 Bearings

Bearings are angles measured from some standard reference direction, usually true or magnetic north. The difference between angles and bearings is that angles are measured from a reference line that can be in any direction but, in modern

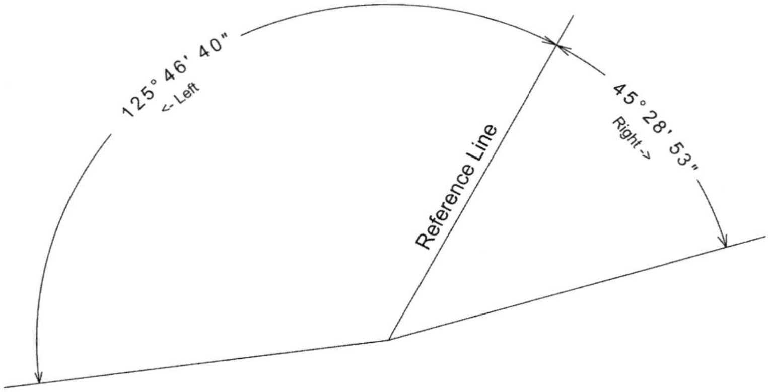

Fig. 4.5 Angles measured left and right from a common line

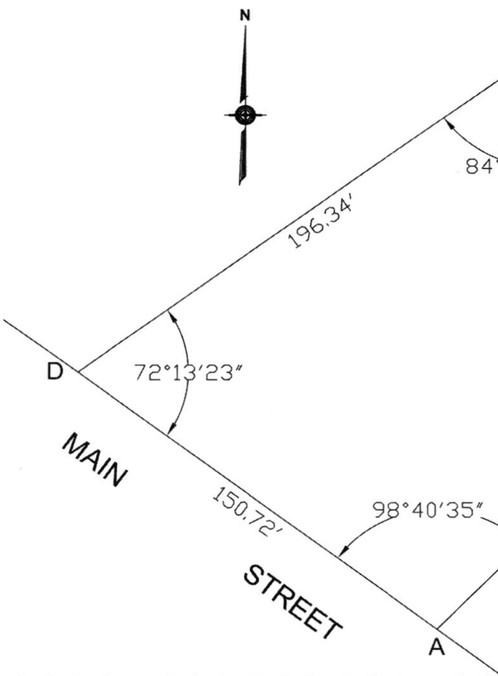

Fig. 4.6 Angles used to describe a parcel of land

surveying, bearings are measured from north or south. An exception to this is that in some very old deeds, drafted in the nineteenth century or earlier, one sometimes finds bearings measured from east or west instead of north.

The difference between angles and bearings is that angles are measured from a reference line that can be in any direction but, in modern surveying, bearings are measured from north or south.

When bearings are used, the circle is divided into **Quadrants**. A quadrant is 1/4 of a circle (90°). Figure 4.7 shows the four quadrants. They are called northeast (NE), southeast (SE), southwest (SW) and northwest (NW).

Figure 4.8 shows the same parcel of land as shown in Fig. 4.6 but with bearings instead of angles. Notice from Figs. 4.8 and 4.7 that bearings are merely angles measured from a common north-south meridian. Bearings beginning with N (north) are measured clockwise (NE) and counterclockwise (NW) from north. Bearings beginning with S (south) are measured clockwise (SW) and counterclockwise (SE) from south. Because bearings are limited to quadrants, no bearing can be greater than 90° (360°/4° = 90°).

Here is another way of thinking about what a bearing represents. Looking again at Fig. 4.8, imagine that you start at the bottom of the page directly below point A and walk in a northerly direction until you reached point A. Imagine you held

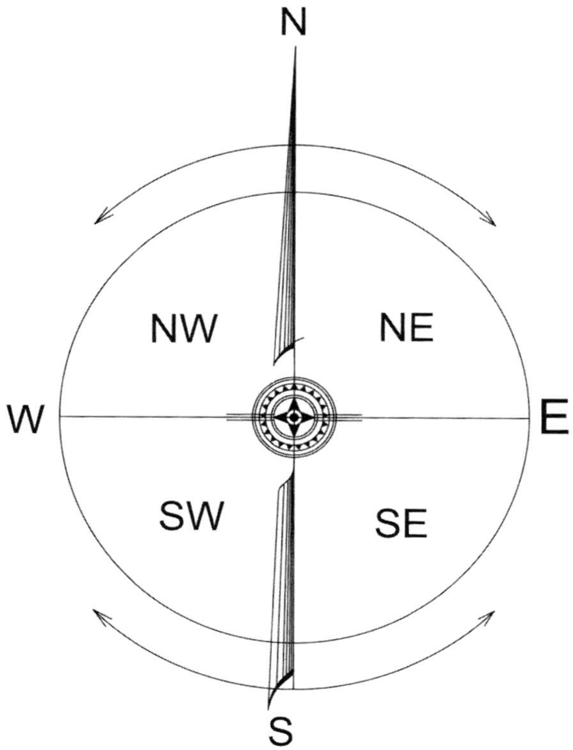

Fig. 4.7 Bearing quadrants

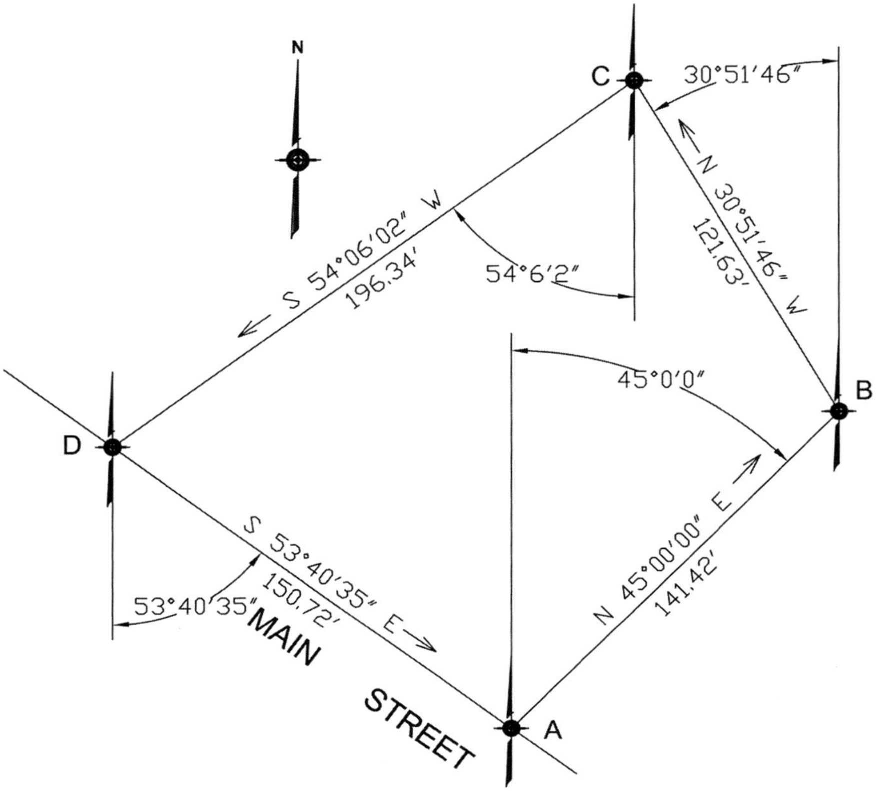

Fig. 4.8 Bearings used to describe a parcel of land

a compass in your hand which had a needle that always pointed north and south. When you arrive at A you make a 45° turn to the right and walk toward point B. You would now be walking N45°E. When you get point B you stop and face north again. You then turn about 30° toward your left and walk in the direction of point C. You would be walking N30°W. When you get to point "C" you stop and face south. Now you would turn 54° to your right and walk to point D. Your walking direction would now be S54°W. When you arrive at point D you stop and face south. You would then turn 53° to your left and walk S53°E to A, the point of beginning.

4.7 Bearing Systems

There are at least four possible interpretations of north and all have been used by surveyors. They are True North, Magnetic North, Grid North and Assumed North. True north is the direction of the meridian that corresponds to the axis of the earth.

It is the line that connects the earth's geographic poles. Magnetic north is the direction that a magnetic compass needle points to when aligned with the earth's magnetic field. Grid north is the direction used on state plane coordinate systems. It corresponds to true north only at the state's central meridian. The deviation from true north becomes greater with increasing distance from the central meridian. Assumed north is an arbitrary north chosen for convenience. Assumed north may have no known relation to true or magnetic north. Let's take a closer look at each of these designations for north.

There are at least four possible interpretations of north and all have been used by surveyors. They are True North, Magnetic North, Grid North and Assumed North.

4.8 True North

True north can be determined by using a transit or theodolite to make an observation of the sun or the star Polaris. It can also be calculated if the surveyor has access to National Geodetic Survey (N.G.S.) control points. N.G.S. (part of the Federal Government) has set and maintains reference points all over the United States. The points are related to a fixed datum such as latitude and longitude and coordinates based on the state plane coordinate system of the state in which the point is located. The points are often bronze disks set in rock outcrops or other permanent features.

The N.G.S. points are often several miles apart. Using conventional surveying techniques it could take several hours or even days to run survey lines to these points. Because of the time involved, it may not be feasible to tie into these points for the purposes of a small survey where the cost of doing so would probably be greater than the cost of performing the survey using a more convenient reference. Many boundary surveyors now have access to GPS receivers. These receivers make finding true north (and grid north) much more accurate, convenient, efficient and cost effective than methods formerly used. As a result of the widespread adoption of GPS, connecting surveys to true north is now quite common.

4.9 Grid North

Grid north is the direction of north as described on a state plane coordinate system. State plane coordinate systems were devised as a means of providing a fixed datum related to latitude and longitude. The coordinate systems have been developed by N.G.S. for each state. State plane coordinate systems are a means of representing the nearly spherical surface of the earth on a flat map. When a spherical surface is projected onto a flat surface a certain amount of distortion occurs. If the area of the map is kept relatively small the distortion is minimized.

For each state plane projection area, a central meridian corresponding to a line of longitude is chosen that is near to the center of the area to be mapped. The central meridian corresponds to true north. As the distance east and west of the central meridian is increased the distortion becomes greater and grid north deviates from true north to a greater extent. This happens because lines of longitude converge at the earth's poles but the grid lines on the state plane coordinate system remain parallel to the central meridian.

Boundary surveys may be related to a state plane coordinate system by running survey lines to N.G.S. points or by using global positioning as described earlier. Connecting a boundary survey into a state plane coordinate system will add a level of redundancy because it has the potential to make the reestablishment of lost monuments easier and more certain. We will discuss GPS methods and state plane coordinates in detail in later sections.

4.10 Magnetic North

Magnetic north is determined by using a magnetic compass. A magnetic compass needle aligns itself with the earth's magnetic field which, in most geographic locations, does not correspond to true north. Approximately 90 % of the earth's magnetic field is generated by the earth's outer core. The relationship between true north and magnetic north varies with the geographic location of the person making the measurement. **Declination** is the term used to describe the angular difference between true north and magnetic north. In 1980 in New Hampshire magnetic north was approximately 15° west of true north. In the Florida panhandle it was 0°. In central Texas it was 8° east and in Oregon it was 20° east. The magnetic declinations for North America in recent times are shown in Fig. 4.9.

> A magnetic compass needle aligns itself with the earth's magnetic field which, in most geographic locations, does not correspond to true north.

The location of magnetic north also varies with time. **Variation** is the term used to describe how declination varies with time. When interpreting old deeds and plans it is necessary to make variation adjustments for historical differences in magnetic north. The National Geophysical Data Center (NOAA) maintains a web site where users can compute the historic declination for locations in the U.S. Having this information available becomes very important when a surveyor needs to locate the boundaries of a compass survey that was performed many years ago.

Although it is probably not an immediate concern for surveyors, the earth's magnetic poles are subject to polar reversals. Every 200,000–500,000 years north becomes south and vice versa. It has been about 750,000 years since the last reversal. It is not known when the next reversal will happen but it is believed that it

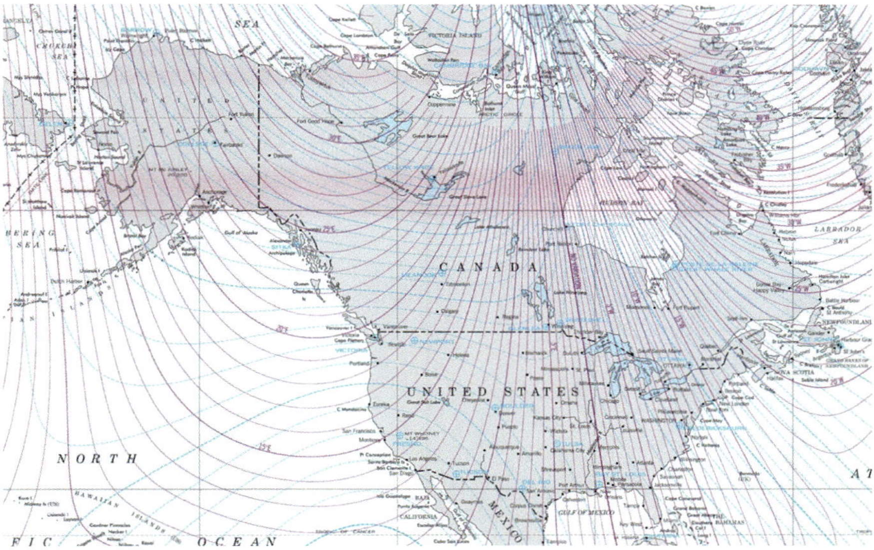

Fig. 4.9 Magnetic declination

might take more than another 1,000 years and perhaps much longer. Magnetic direction is also influenced by local attraction and magnetic anomalies occurring in the earth. Magnetic attraction can be caused by steel or iron objects such as vehicles or metal buildings in the vicinity of the compass. Magnetic fields caused by nearby power lines can also affect the compass.

A magnetic compass in the center of a surveying transit can be seen in Fig. 4.10. The north direction of the compass is in line with the telescope axis so that the magnetic direction of a line can be read from the compass simply by sighting an object through the telescope and reading the compass dial. The compass needle contains a jewel bearing which reduces friction and wear. To prevent damage and wear while the transit is transported, the compass needle should be lifted off its bearing when it is not being used. Although not visible in the image, most surveying transits have a screw or other device which raises and locks the needle so that it is not constantly resting on the bearing.

The needle of a magnetic compass is subject to **Magnetic Dip** as shown in Fig. 4.11. Because the earth's magnetic pole is not on the horizon, but inside the earth, the magnetic force causes the needle to tilt downward toward the ground (in the northern hemisphere) as it tries to align itself with the magnetic lines of force. As the compass gets closer to the magnetic pole (higher latitudes), the amount of dip becomes greater. The needle of a magnetic compass can be weighted on one end to help offset the vertical component of the magnetic force and cause the needle to remain horizontal, or nearly so. On surveying compasses, the weight is often a bit of wire wound around the compass needle.

Fig. 4.10 Magnetic compass

Small magnetic compasses are not particularly accurate or precise by the standards of modern surveying. For example, if a surveying compass can be read to 1/10 of a degree and a theodolite can measure an angle to 1 second of arc, the theodolite will be 600 times more precise than the compass. In spite of these limitations, because of economic or other considerations, a compass may still be a surveyor's only available source of direction. Normally, a compass will be used to

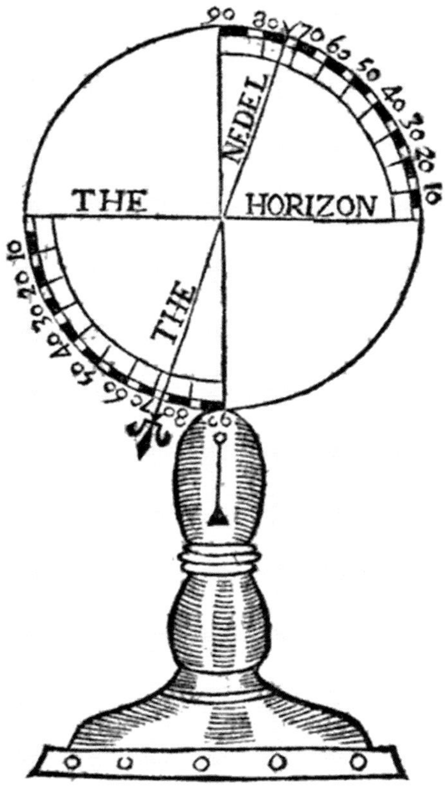

Fig. 4.11 Magnetic dip

obtain the bearing of one boundary or traverse line. A theodolite or total station will then be used to measure precise angles from this line to all other lines in the survey. Using this method, the boundary measurements will be precise. Only the orientation of the parcel will be limited to the accuracy of the compass and the stability of the magnetic field. This latter deficiency can mitigated by setting permanent boundary markers on the ground to orient the parcel.

During the nineteenth century, and earlier, surveyors were limited to using a compass to obtain a magnetic bearing for the survey, or making astronomical observations to determine true north. Making astronomical observations is time consuming and it is sometimes cloudy. As a result, determining true north was rarely used for small surveys. For surveys that did not require precise measurements, boundaries were measured using only a magnetic compass and a chain. Because the method is imprecise, this technique is rarely used today for boundary determination. However, modern surveyors performing retracement surveys of these ancient surveys must take these factors into consideration when locating boundaries so they must understand the issues associated with magnetic measurements.

4.11 Bearing System Confusion

Bearing systems are often a great source of confusion among those not experienced in working with them. As we have seen, a surveyor has a number of choices regarding which north to use. The choice could be true north, magnetic north, grid north or some arbitrary north that may have no known relation to an actual north. Different surveyors surveying nearby or adjacent properties may make different choices about what to use as north. In small surveys, such as house lot surveys, the choices are primarily driven by economic considerations. The cost of determining true north or grid north may be greater than the cost of the survey itself. It is no surprise then that there are many plans on record which are on different bearing systems. In some cases, plans exist describing abutting properties which share common boundary lines and these plans show different bearings for the common lines.

How easily this happens in the real world may be illustrated by an example. Suppose a surveyor has found a plan on record of a small lot on a magnetic bearing system. The surveyor has also found a plan showing a state highway layout based on true north. The property to be surveyed lies between the highway and the previously surveyed lot. If the surveyor uses the highway system, the common boundary with the lot will have bearings that are different from those shown on the lot plan. If the surveyor uses the magnetic bearing system, the common boundary between the lot to be surveyed and the highway will be on different bearing systems. In such cases the surveyor will have to decide what bearing system to use. Either decision will result in bearing system inconsistencies.

Another problem is that many plans on record do not give information about what bearing system was used. Often a plan will show a north arrow but no information is provided describing which north the arrow represents. From a practical point of view it doesn't always make a great deal of difference to a surveyor which bearing system a previous surveyor has used. Use of inconsistent or arbitrary bearing systems is not necessarily a fatal flaw in the survey if enough common permanent markers are called for on the plans and deeds and as long as the markers are still in existence on the ground. It is possible to equate the different systems by mathematically rotating one of the systems to match the other system. The reader will understand that one purpose of the bearings and distances which describe a parcel of land is to define a rigid geometric figure. In cases where bearings cannot be relied upon to orient the parcel with respect to some reproducible meridian, physical evidence called for in the description and found on the ground can be used to locate and orient the parcel.

Many plans on record do not give information about what bearing system was used. Often a plan will show a north arrow but no information is provided describing which north the arrow represents.

It would be much simpler and more convenient if all plans were on the same system and that system was easily reproducible on the ground. The decreasing cost of global positioning equipment is beginning to make it practical for surveyors to relate their surveys to fixed datums, such as state plane coordinate systems, so that in future it will be more common to have surveys related to the same bearing and coordinate system. Legislation or municipalities may require that all plans submitted to them be on the state plane coordinate system.

Given the millions of plans and deeds already on record, surveyors will have no choice but to continue to work with inconsistent bearing systems. When reading deeds and plans, researchers must be aware that bearings given on a particular plan or deed may not correspond with bearings given in another plan or deed, even though the plans or deeds are describing the same line. As we have seen, the fact that common lines may be described by different bearings on different plans does not necessarily mean that either of the plans is incorrect. As veteran boundary surveyors already know, it is precisely these inconsistencies that make boundary surveying so interesting and challenging.

4.12 Bearings Versus Angles

Most modern surveyors use bearings. If bearings are used it is always possible to calculate the angles from the bearings, but unless one bearing is given, angles cannot be used to calculate bearings. If a surveyor makes a mistake in writing an angle on a plan it can be more difficult to find the source of the error than if bearings had been used. In surveying, an angle relates a line to a previous line whereas a bearing relates the line to a meridian. If bearings are related to true north or other known meridian the directions are fixed on the earth and can be always be reproduced. In contrast, angles only give the relationship between the property lines themselves. In other words, angles allow the description of a rigid geometric figure but they do not orient that figure on the ground as do bearings.

4.13 Reversing Bearing Directions

Any line described by a bearing can have two directions exactly 180° apart, depending which way you are traveling on the line. Consider the line running from A to B in Fig. 4.8. If you are going from A to B the bearing is N45° 00′ 00″E. If you are going from B to A the bearing would be S45° 00′ 00″W. If this doesn't make sense to you look at the description of quadrants in Fig. 4.7. The quadrant opposite NE is SW. Reversing bearings is easy; all you have to do is exchange North for South and East for West or the other way around. The numbers do not change, only the quadrants change.

Any line described by a bearing can have two directions exactly 180° apart, depending which way you are traveling on the line.

Sometimes, when a legal description is drafted from a plan, the bearing directions in the description are wrong because the wrong quadrant is used. How easily this can happen may be illustrated by considering the subdivision plan shown in Fig. 4.12. The plan shows two lots with a common boundary. The common boundary has a bearing of S 07° 29′ 25″E. A description of lot 1 can be created by simply copying the bearings from the plan. No matter what point of beginning is chosen, all of the bearings proceed in a clockwise direction around the lot. A problem arises, however, when describing lot 2. The bearing for the common boundary between lots 1 and 2 cannot be copied from the plan. The bearing must be changed to N 07° 29′ 25″W. Suppose the person writing the deed description for lot 2 chose point A for the point of beginning and wrote the following description copied directly from the plan: "…thence S82° 30′ 35″W, 200.00 feet to a corner; thence S07° 29′ 25″E, 200.00 feet to a corner; thence N82° 30′ 35″E, 200.00 feet…" The result is shown on the plan in Fig. 4.12 with dashed lines. It is apparent that the description does not correctly describe the lot. Most boundary plans do not show

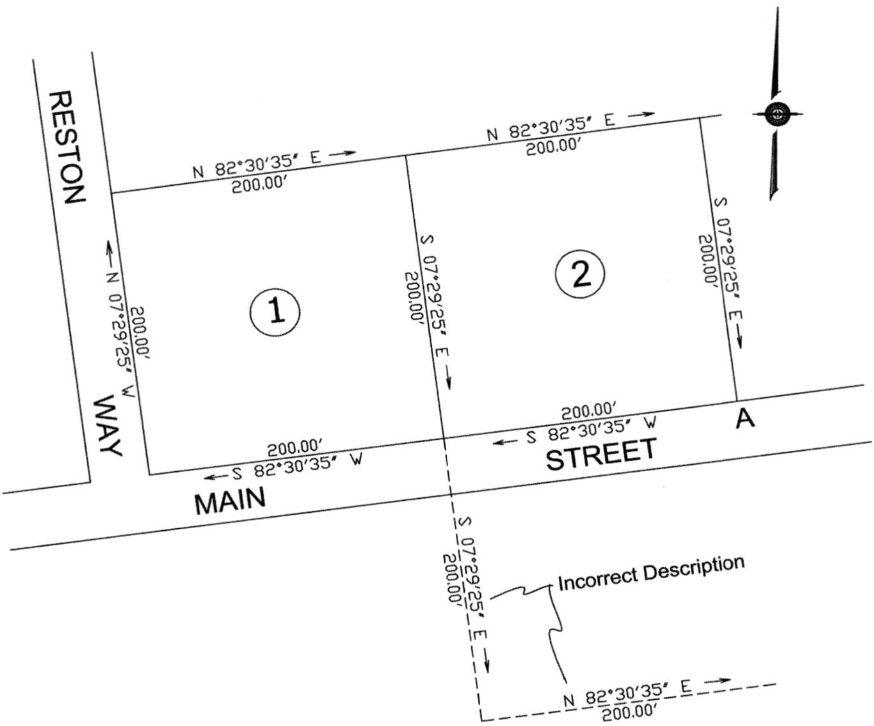

Fig. 4.12 The effect of an incorrect quadrant

two bearings on common lot lines so a person drafting a legal description (usually an attorney) must be careful to choose the correct quadrants for the bearings.

The above example is a common mistake. In a simple case, such as the one illustrated, the error would probably be obvious to the person reading the deed description, particularly if that person had a copy of the plan. However, this type of error may be much more difficult to locate in a boundary survey having many boundary lines where there is no plan available to guide the reader.

4.14 Cartesian Coordinates

In boundary surveying, surveys are usually plotted and defined using a Cartesian coordinate system. A two dimensional Cartesian coordinate system is a rectangular coordinate system where the axes are 90° apart. Any point can be located using just two numbers. These numbers are called **Coordinates**. In most rectangular coordinate systems the axes are labeled X for the horizontal axis and Y for the vertical axis. Values increase up and to the right. In boundary surveying it is customary to label the axes as **Latitude** and **Departure**. Latitude (Y) is measured north and south and Departure (X) is measured east and west.

> In boundary surveying, surveys are usually plotted and defined using a Cartesian coordinate system.

Refer to Fig. 4.13, which shows two points plotted on a coordinate system. If it were necessary to show elevations, a third or Z coordinate would be shown. In the example, the coordinates for point 1 are labeled North 5040.000 and East 4920.000. Once the coordinates of two points are known, it is possible to calculate the bearing and distance between them. In our example, the latitude and departure between points 1 and 2 is determined by subtracting one coordinate from the other. For example:

$$Latitude = 5140.000 - 5040.000 = 100.000$$
$$Departure = 5020.000 - 4920.000 = 100.000$$

Once we know the latitude and departure of a line we can use trigonometry to calculate the bearing of the line and the length of the line. In this case the angular solution is straightforward. Both the latitude and departure are the same so the line running from point 1 to 2 must be N45°E. The method of using the coordinates of two points to calculate the bearing and distance between them is called an **inverse**. The appendix has examples of using trigonometry to calculate inverses.

Some coordinate systems are arbitrary and some are not. In our example, we chose an arbitrary coordinate to begin with at the lower left corner: N4,980 and E4,860. The starting coordinate values do not matter because all calculations are made using the differences between the coordinates. The coordinates chosen for our initial point really do not matter. When selecting a starting coordinate, most

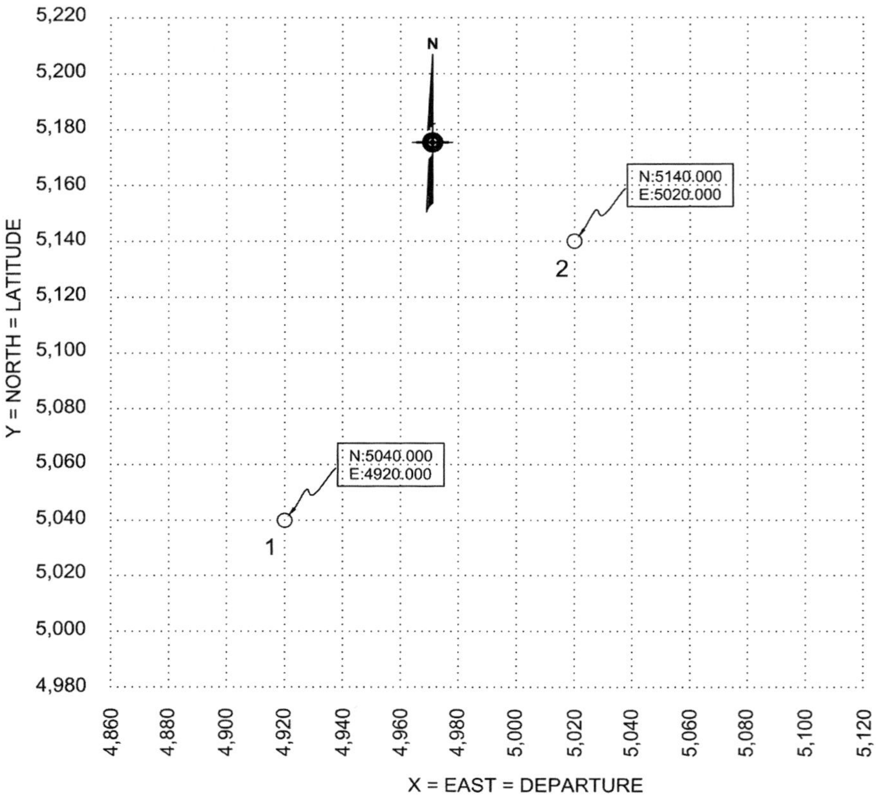

Fig. 4.13 Cartesian coordinate system

surveyors will choose a large enough value such that the coordinates never go below zero. This avoids having to work with points having negative coordinate values.

If we choose to tie a survey into a state plane coordinate system, we will have to use coordinate values adopted by the state. We have already discussed the other issue to consider when choosing coordinate systems, and that is the bearing system. If we decide to use magnetic north, the Y axis will need to be oriented in line with magnetic north. The same would hold for true north. If we are on a state plane system, that decision will have already been made for us.

4.15 The Concept of a Property Line Closure

Surveyors use the mathematical concept of Closure as an indication of the correctness of a boundary description. Closures are very important because if a boundary description does not close, it may be difficult to establish the boundaries on the ground with a reasonable degree of certainty.

Surveyors use the mathematical concept of Closure as an indication of the correctness of a boundary description.

The concept of a closure is relatively straightforward and may be illustrated by a simple example. Imagine that Fig. 4.14 is a small rectangular garden with a fence running around its perimeter. There is a break in the fence so the gardener can enter and exit the garden. The break is shown in the bottom left corner of the image. Let's assume that you mentioned the garden to a friend who wants to create a similarly sized garden on her property. She lives out of town so she asks if you know of someone who could measure the garden for her and send her a written note with the dimensions. You ask the gardener to do it and he takes a measuring tape and begins measuring at the point labeled **Point of Beginning**. He continues to measure along the fence in the direction shown by the arrows and stops at the point labeled "End Point". If you add up the lengths of the 4 sides that he measured, they will not give the actual size of the garden, because he neglected to measure across the two foot gap in the lower left corner of the garden. In surveying, the omission in his measurements would be called an **Error of Closure**. In our example, the perimeter of the garden would actually be 360 feet, but he only measured 358 feet. The measurements therefore have an error of 2 feet.

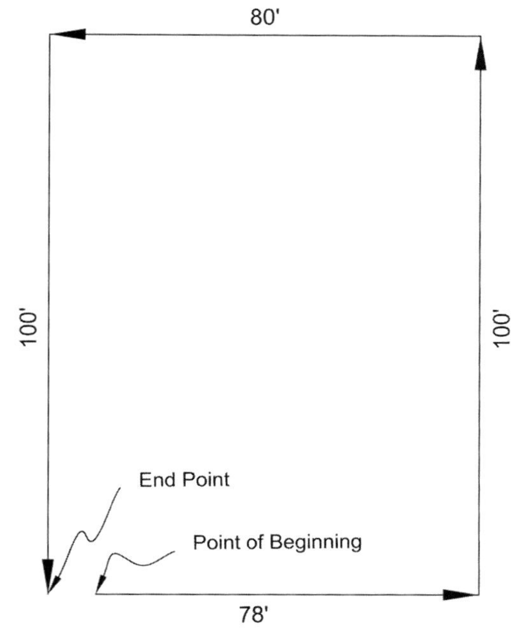

Fig. 4.14 Garden Fence error of closure

Now, imagine that the gardener didn't notice the error and he sent a handwritten note to your friend with the four dimensions. He also makes a note that the garden is a rectangle (in a rectangle all of the interior angles are 90°). If your friend decided to draw the garden to scale on a piece of graph paper using the gardener's dimensions, she would soon find that there was a 2 foot gap. She would also discover that without additional information, there would be no way to tell which of the two sides contained the error. Remember, she doesn't have a sketch; she only has a note containing four dimensions. In a rectangle, the opposite sides have to be the same length. So, in our example, the sides containing the error could either be 78 or 80 feet. She would know the sides must either be 78 or 80 feet, but she would not be able to determine which dimension was correct.

This is the same sort of dilemma that one sometimes runs into when trying to interpret a deed description that contains errors. The description may have dimensions and directions for each boundary, but when they are plotted, there is a gap between the beginning and end points. Sometimes, figuring out where the error was made can be quite difficult or even impossible. Unlike your friend, who could just send the gardener a note asking him to recheck his measurements, a surveyor often has no way of knowing where the deed dimensions came from. Of course, the error in our example is huge, at least by boundary surveying standards, so let us look at a more realistic example.

Consider Fig. 4.15 which shows the property boundaries of a parcel of land as established on the ground by a survey. The boundaries shown in the figure are the same boundaries shown in Fig. 4.8 but the bearing and distance along Main Street is slightly different because small errors in measurements were made when the survey was performed.

If you were to start at the property corner labeled "A", which we will call the Point of Beginning, and precisely measure counterclockwise around each boundary line using the bearings and distances shown, you would arrive at a point very near the point of beginning but not exactly at the point of beginning. It is important to accept the fact that even the most precise survey will always have small errors in measuring angles and distances, so the end point will rarely correspond exactly to the beginning point. Even if a survey did close perfectly on paper, there are probably still measurement errors—the errors have probably cancelled each other out.

The sum of the errors in a survey will give an error of closure. A similar situation exists when using GPS except that, because there are no angles and distances measured, the points themselves will contain uncertainties in their locations.

The error of closure is described by a bearing and a distance which represents the line that would have to connect the end point with the point of beginning in order for the figure to close perfectly. The error of closure is shown in Fig. 4.15 as N14° 10′ 20″E, 0.05 feet (about 5/8 of an inch). The error has been exaggerated in the drawing for purposes of illustration. The reader can see that the error of closure represents an additional line in the property boundaries.

Although the error of closure consists of a single bearing and distance, it is important to understand that the error is usually the result of the accumulation of the errors which exist in each of the bearings and distances comprising the perimeter

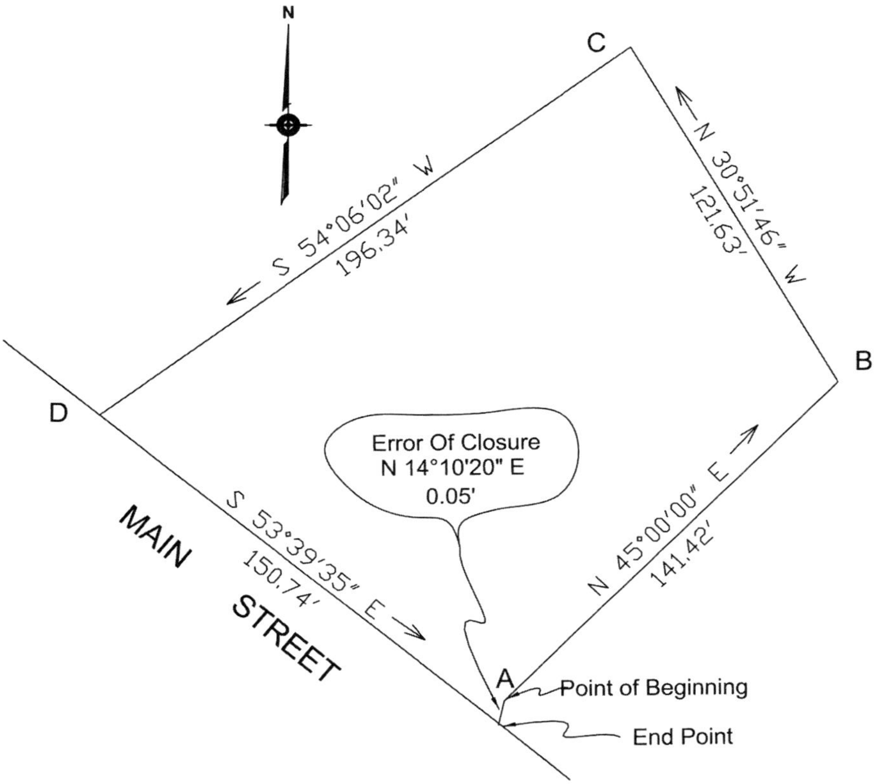

Fig. 4.15 Example of an error of closure

of the property being described. For example, in Fig. 4.15 each of the four boundary lines could contain small errors. Sometimes these errors will accumulate and sometimes the errors will tend to cancel each other. The resultant error of closure represents the sum (actually the vector sum) of the all of the errors. Large errors of closure may be indicative of sloppy measurement techniques, a poorly calibrated instrument or tape or a blunder. If measurements are recorded manually, a large error of closure may result from improperly recording readings or measurements.

Sometimes, the error of closure may provide a clue as to where the error may have occurred, particularly if the error is large. For example, in Fig. 4.15, if the bearing of the error of closure were N54°E, there is a good chance that the length of line which has a bearing of S54° 06′ 02″W was measured incorrectly. Alternatively, either or both of the angles at points A or B may have errors.

In the chapter on calculations we will see that the concept of closure also applies to traverses. (Traverses are the actual lines which the surveyor measures. Traverse lines are usually not the same as property lines. This will be described in detail later on.)

4.16 Error of Closure Standards

Some states and governing authorities have rules establishing classes of surveys or limits on errors of closure or, when using GPS, errors in position. An error of closure can be represented as a ratio of the magnitude of the error divided into the sum of the length of all of the boundary lines. For example, in Fig. 4.15 the sum of the length of the boundaries is 610.13 feet. That is the total distance we would have to measure if we began at point A and measured each boundary line consecutively until we arrived again at point A. If the error of closure were 0.05 feet, then the ratio would be expressed as 1 foot in 12,203 feet (610.13/0.05) or 1 part in 12,203 parts. Another way of thinking about it is if you were to measure 12,203 feet (about 2 miles) you would be off by 1 foot.

Smaller ratios are indicative of greater errors. For example if the error in the above example were 0.09 feet the error of closure would be 1 in 6,780 (610.11/0.09). For many boundary surveys an error of closure of 1 in 10,000 to 1 in 15,000 is considered to be the minimum acceptable closure. The American Congress on Surveying and Mapping (ACSM) standard for measurements which control land boundaries has a minimum requirement of 1 in 15,000. A closure of 1 in 10,000 corresponds to 0.01 feet in 100 feet, which is the smallest graduation available on a steel surveyor's tape. When using total stations, errors of closure of 1 in 50,000 or even 1 in 100,000 are not uncommon. One in 50,000 roughly equates to a one foot error in 10 miles.

4.17 Understanding the Accuracy of Measurements

In some cases, when we specify a numerical value, the number is exact. For example, my dog has two puppies, or $5 + 5 = 10$. Clearly, a dog could not have 2.4 or 2.5 puppies. There is no uncertainty in the value. There is also no uncertainty in the addition of two integers: $5 + 5$ cannot equal 10.1. In contrast, when we make a measurement, there is always some uncertainty in the measurement. Boundary surveying involves making measurements and the measurements will always contain uncertainties. Surveyors need to understand these uncertainties and know what level of confidence they have in their measurements. Every measurement must include an estimate of the confidence in the measurement. For example, suppose we use a steel tape to measure a distance between two points that are exactly 100 feet apart, and we can only read the tape to the nearest hundredth of a foot (0.01), we might say the distance is 100 feet plus or minus 0.01 feet. The "plus or minus 0.01 feet" tells us something about the level of confidence we have in the measurement. We could also use a tape graduated in tenths of a foot (0.1) to make the same measurement. If we were limited to reading the tape to the nearest tenth, we would have much less confidence than in the previous example that our measurement was actually 100 feet.

The goal, when measuring property is to know the true value of the measurements. It turns out that knowing the true value of a measurement is not always a simple task. We have seen that the survey foot is used in the majority of jurisdictions in the United States. The survey foot is specified as a certain fraction of a meter, which has an internationally defined length. In surveying, when we speak of the true value of a measurement, we are relating the measurement to a known and verifiable standard. If we are making measurements with a steel tape, the tape must be calibrated against the standard. If we are making measurements with a total station with an electronic distance measuring device (EDM), then the EDM system must be calibrated to the standard. In actuality, all measurement devices and methods will contain some error. All we can do is to be sure that our measurement instruments are well calibrated and that we are using proven measurement techniques so that our measurements will as close to their true value as possible.

The goal, when measuring property is to know the true value of the measurements.

4.18 The Accuracy of Measurements Using GPS

The error of closure we have been discussing up until now is based on traditional surveying techniques where linear measurements have been used, for example measurements taken with a tape or total station. These types of measurements are based on measuring straight lines between fixed points. If we are using GPS, the measurements are less dependent on the distance between points and more dependent upon statistical probabilities of the location of a point. These different methods for expressing uncertainty in measurements and accuracy standards has made it more difficult to compare the accuracy and uncertainty of points established using linear techniques with those established using GPS.

If we are using GPS, the measurements are less dependent on the distance between points and more dependent upon statistical probabilities of the location of a point.

When working with GPS locations, the horizontal accuracy is expressed using a positional uncertainty called the **Circle of Uncertainty**. The accuracy of a particular point is based on the 95 % probability of the point being within a circle having a specific radius.[1] An example may help to clarify this. Assume we are using a

[1] Based on a 95 % confidence level. See Standards and Guidelines for Land Surveying using Global Positioning System Methods, August 6, 2004, The Survey Advisory Board for Washington State Department of Natural Resources.

GPS receiver with a horizontal accuracy of 1 cm + 1 ppm. The circle of uncertainty would mean that if we drew a circle with a radius of 1 cm around our point and made 100 measurements, 95 of the measurements would be inside the circle. However, we can never really know with certainty than any particular point will be an exact distance from the center of our circle. We can see from this example that evaluating the accuracy of points measured with GPS is fundamentally different than it would be if the points were measured in a conventional traverse survey where angles and distances were measured with a total station.

The same GPS circle of uncertainty will apply equally to a house lot where the corners are 100 feet apart and a large parcel where the corners are 3,000 feet apart. In the case of a house lot, an accuracy of 1 cm might not be acceptable, but in the case of the large parcel it might. This is because surveyors need to be able to locate property corners with a specific accuracy. For example, if our standard is 1 in 10,000, and the boundary distance of our house lot is 100 feet, we must locate the corner within 0.01 feet (1 foot in 10,000 feet is 0.01 feet in 100 feet). If the GPS receiver accuracy is 1 cm this translates to 0.03 feet. The GPS will therefore not be accurate enough for establishing the lot corner. On the other hand if our boundary line is 3,000 feet long, based on our 1 in 10,000 requirement our linear tolerance is now 0.30 feet. Because the GPS accuracy of 0.03 feet has not changed, we can expect 95 % of our points to be within 0.03 feet so our accuracy is 10 times better than required. Looked at another way, using the GPS receiver, the points would need to be at least 330 feet apart to meet our standard. Of course, it is possible for regulatory agencies to have different positional tolerances for GPS and linear techniques.

When considering the purchase or use of GPS receivers it is important to compare positional accuracies at the same certainty level. Manufactures may specify certainty levels of 68.3 % rather than 95 %. In our previous example using a circle of uncertainty with a radius of 1 cm, if we used the manufacturer's probability, only 68 of the points measured would be in the circle, so we would have considerably less confidence in the measurement than in the case of 95 % probability. In addition to the circle of uncertainty, GPS manufacturers also specify an additional ppm (parts per million) figure. This simply means that the stated error must be increased for each part per million of baseline length—the distance between the receiver and base station. The ppm adjustment is usually quite small, 1 ppm being about 0.01 feet in two miles.

In our first example, we used significant digits such as 0.1 and 0.01 as an indication of uncertainty, and it is true that the number of decimal places shown is an indication of uncertainty. However, it is important to understand that this rule can lead to undesirable results in surveying. For example, a surveyor may locate the middle of a shallow stream which is 20 feet wide by standing in the middle of the stream with a prism pole. The data collector will record a location to the nearest hundredth of a foot even though the person holding the pole may not be able to tell where the middle of the stream is within a foot or more. Although the location of the prism pole is accurate within 0.01 feet, the location of the stream is only accurate to a foot. In the end, we are really only concerned with the location of the stream, not the pole, so it will be up to the surveyor to annotate the report or plan so as not to overstate the accuracy of the measurement. When recording data in field books was common, this was less of a problem because the distance could be recorded with the correct precision.

4.19 The Difference Between Precision and Accuracy

The terms precision and accuracy are often mistakenly used to mean the same thing when they actually have quite different meanings. **Precision** is the degree of refinement with which a measurement is made. It has to do with the closeness or agreement of a series of repeated measurements. For example suppose we have a steel tape that is graduated in hundredths of a foot (0.01′). We want to know the property line distance between two fixed concrete monuments. We make three measurements with the tape and read: 100.03, 100.00 and 99.98. Suppose we then made a second set of measurements and get 99.99, 100.01 and 100.00. The second set of measurements is more precise than the first because the deviation between measurements in the second set is less than in the first set.

Accuracy has to do with the true or correct value of a measurement. In the preceding example suppose the distance between the two bounds was measured with a faulty tape that was missing exactly one foot. Imagine that the tape had been broken near the middle and someone mistakenly cut a foot off the tape when they spliced it back together. When measuring between the bounds the tape would read 101 feet not 100 feet. Suppose we then used the faulty tape and made the following series of measurements: 100.99, 101.01 and 101.00. Because the true distance between the bounds is 100.00 feet not 101.00 feet we might say that these measurements were reasonably precise but not very accurate.

As the previous example shows, precision doesn't have much value without accuracy. Surveyors need to be able to make very precise measurements but those measurements must also be as close to their true value as possible.

Chapter 5
Surveying Instruments

Surveying instruments are used for measuring angles and, when using total stations, they also measure distances. Transits and theodolites were formerly used for measuring angles. These were mechanical and optical instruments which measured horizontal and vertical angles. Now, digital **Total Stations** are used almost exclusively. Total stations are theodolites which include electronics capable of measuring distances. They also have onboard electronic data collection and/or provision to connect to an external data collector. With tracking total stations, the telescope is able to automatically track the person making the locations. Robotic total stations eliminate the need for a person to be at the instrument in order to make measurements, because the telescope can be controlled remotely. The term **Instrument** is generic in that it is often used to refer to transits, theodolites and total stations.

Survey grade GPS is also widely used for boundary surveying. For many surveys, GPS is preferable to total stations because of its speed and the potential for high accuracy over long distances. However, GPS accuracy is affected by a number of environmental factors, limiting its use under certain conditions, so GPS has yet to replace total stations for many types of work and environmental conditions. Often, both GPS and conventional surveying techniques can be combined thereby taking advantage of the strengths of each technology. In this book, the term *conventional surveying techniques* is used to describe using total stations to make linear measurements, as opposed to using GPS.

GPS accuracy is affected by a number of environmental factors, limiting its use under certain conditions, so GPS has yet to replace total stations for many types of work and environmental conditions.

P. Gay, *Practical Boundary Surveying*, DOI 10.1007/978-3-319-07158-9_5

5.1 Transits and Theodolites

Angles were traditionally measured using optical instruments such as **Transits** and **Theodolites**. These instruments typically have a telescope of 20–30 power magnification. The telescope is fitted with a **Reticle**. A reticle is a set of cross hairs similar to that found on a rifle scope. The reticle consists of horizontal and vertical lines, crossing in the center of the telescope. The reticle allows a target to be precisely sighted both horizontally and vertically.

Surveying instruments also have a short set of horizontal **Stadia** hairs, above and below the horizontal cross hair, which allow **Rangefinding**. Rangefinding is the ability to optically measure the distance from the instrument to a target. Cross hairs and stadia hairs are shown in Fig. 5.1. On instruments found in the United States, the stadia hairs are spaced so the vertical distance between them is exactly 1 foot at 100 feet. A graduated vertical rod, as shown in the image, can be used to read off the approximate distance from the instrument to the rod. Most rods are graduated in hundredths of a foot (0.01′) so distances measured using stadia are accurate to about 1 foot. In the image in Fig. 5.1 the distance would be about 21 feet. Stadia measurement was handy for rough measurements before electronic measuring devices became common but it is rarely used today.

Transits and theodolites have clamps and fine adjustment screws, called **Tangent Screws**, which allow very precise positioning of the telescope in both the horizontal and vertical axes. A tangent screw and clamp are shown in Fig. 5.2. Each axis has its own clamp and adjustment screw. When sighting a target, the instrument is manually turned and aimed at the target. This is best accomplished by visually sighting over the telescope with both eyes open, not by looking through the telescope.

Some telescopes have crude metal sights or collimators on the outside of the telescope for this purpose. The Wild T1A shown in Fig. 5.3 has a mechanical peep sight just above the eyepiece. The front sight, which is a bit difficult to see in the image, is a short black post just above the objective lens. (The feature mounted

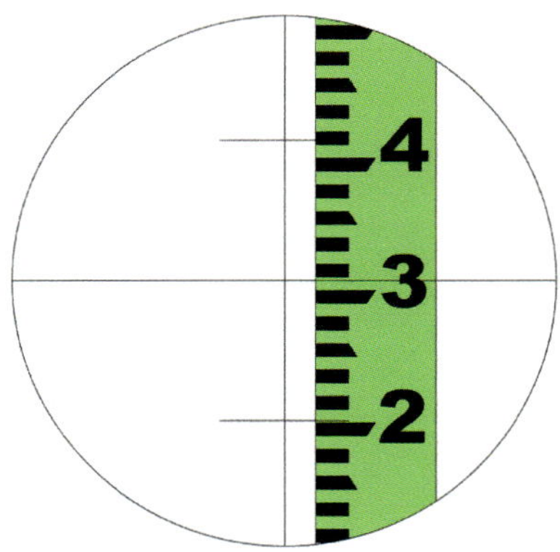

Fig. 5.1 A reticle with stadia hairs

Fig. 5.2 Tangent screw and clamp

to the top of the instrument standard on the right side, which looks like a small telescope, is actually a compass.) When the telescope is roughly aimed at the target, the horizontal clamp is tightened. The target would then be sighted through the telescope where fine adjustments are made using the tangent screw. A novice will often attempt to initially sight the target by looking through the telescope. This is rather difficult, particularly for a novice, because the high telescope magnification results in a very small field of view which makes it challenging to find the target. Additionally, only one eye can be used when looking through the telescope. The user cannot take advantage of binocular vision to help spot the target. Studies have shown that two eyes are better than one for such visual tasks. Finding the target can be demanding when it is in dense woods where the light is dim or when there is only a tiny hole through leaves or underbrush where the target is visible.

Transits and theodolites are capable of measuring horizontal and vertical angles with a high degree of precision. The instruments have horizontal and vertical circles which are used to measure the horizontal and vertical angles. The construction of these circles varies depending on the type of instrument. Transits usually have

Fig. 5.3 Wild T1A optical theodolite with peepsight

metal circles with **Verniers**. A vernier is a specially divided scale which allows more precise reading of the arc. Transits can be read to about 1 min or 30 seconds of arc or even a little less. Theodolites have glass circles with markings and verniers etched on the glass. Unlike a transit where the circles are more or less open to the environment, on a theodolite the circles are internal. Theodolite circles, on the more precise instruments, can be read to 1 second of arc or less. Modern total stations have optical encoders on the glass scales enabling angles to be read digitally.

As noted, the most precise instruments are capable of resolving an angle to 1 second of arc or less. To put this into perspective, 1 second of arc at 1,000 feet is about 0.005 feet, less than 1/16 of an inch! The Wild T1A theodolite shown in Fig. 5.3 reads to 20 seconds of arc. Although these instruments have been largely replaced by total stations, which can also measure distances, they were made in Switzerland and are of superb quality.

5.2 Total Stations

Total Stations are theodolites which have built in **Electronic Distance Measuring Devices (EDM)**. These devices emit light signals at varying frequencies. The light signals are reflected back to the instrument from a prism and the EDM is able to accurately measure the distance between it and the prism. On some of the earlier total stations, the EDM was mounted above the telescope. This allowed an EDM to be retrofitted to some older theodolites. With this design, when the telescope was pointed downwards, the EDM would move forward of the telescope axis. When the telescope was pointed upwards, the EDM would move backward from the telescope axis. This meant that distances measured with the EDM had to be adjusted for the offset. The greater the telescope deviated from horizontal, the greater the adjustment. A top mounted EDM is shown in Fig. 5.4 where it can be seen just above the telescope eyepiece. All new total stations place the EDM inside the telescope so the EDM is coaxial with the line of sight, thereby eliminating the need for offset compensation.

To measure a distance with a total station, the telescope is precisely pointed at the prism and a button on the total station or data collector is pressed to initiate the measurement. When a data collector is used to store the data, the information is transmitted from the total station to the data collector. Because the prism is rarely at exactly the same elevation as the total station, the distance measured is always a slope distance. When the total station sends data to the data collector, it is therefore necessary for the total station to transmit three pieces of information: the horizontal angle, the slope distance and the vertical angle. When we discussed measuring distances with a steel tape it was shown that if a slope distance and vertical angle were known it would be possible to calculate a horizontal distance. Most total stations have onboard software which converts the slope distance to a horizontal distance. Recall that in boundary surveying we are primarily concerned with horizontal distances, not slope distances. With an EDM it is possible to measure a mile to an accuracy of a few hundredths of a foot or less in a matter of 1 to 5 seconds. When using an EDM it is necessary to see between the points being measured—the points must be intervisible. This is usually not a problem because in order to measure an angle, the points must be visible through the telescope. Nevertheless, in some situations, such as when the wind blows a tree leaf online, or when a small twig is exactly online, it will be possible to sight the target but the EDM will not be able to measure the distance until the obstruction is removed.

Fig. 5.4 Total station with data collector—the data collector is attached to the tripod and connected to the total station with a communications cable. Note the top mounted EDM

5.3 Reflectorless Total Stations

Many total stations have **Reflectorless** capability. A reflectorless total station does not require a prism to reflect the light back to the EDM. The emitter in a reflectorless EDM is much more powerful than a conventional EDM which relies upon a prism to reflect light back to the total station. So, even when measuring to a relatively dull surface such as a tree, some of the light will be reflected back to the EDM. The range is dependent on the type of surface and color of the surface which reflects the light. Because most surfaces are much less efficient at reflecting light than a prism, the range of a reflectorless EDM will be considerably less than when using a prism. Depending on the power of the EDM and the quality of the reflective surface, ranges of roughly 500 feet are possible. In contrast, when measuring to a prism, the EDM may measure a mile or more.

A reflectorless total station does not require a prism to reflect the light back to the EDM.

Reflectorless EDMs are often used for locating buildings and other features which cannot be conveniently measured using a prism. For example, the top of a tower or other inaccessible feature is easily measured with reflectorless total station. With a conventional EDM it would be necessary to place a prism on or next to the object being measured. For many features, this would be inconvenient or impossible. Boundary surveys are often required because a neighbor is disputing a boundary location. In cases where it is necessary to measure objects on a property where the owner is hostile or uncooperative, a reflectorless EDM often allows the measurements to be made without the necessity of obtaining a court order or injunction to enter upon the property. Even if a surveyor is allowed on abutting property by statute, reasonable notice to the property owner is normally required. Sending registered letters takes time and the letter may not be accepted. With a reflectorless EDM it is only necessary to set up the total station on a location near but not on the inaccessible property and take measurements to the features which need to be located. Not all objects can be located this way but many can.

Some objects, such as the corner of a house may be difficult to measure accurately with a prism. We will discuss prism offsets later in this chapter. For now it is sufficient to know that it is a distance correction which must be applied to distances measured using a prism. The offset for a particular prism is based on the hinge point of the prism or the plumb line of the prism holder. The plumb line is not always at the back of the prism so if a distance needs to be measured to the corner of a building, the distance will be incorrect because the plumb line of the prism will not be at the corner of the building. The reflectorless total station solves this problem because the light will be reflected from the exact corner of the building.

Reflectorless total stations do require extra caution on the part of the instrument person. The EDM doesn't care what it is measuring to. As long as the target is within range it will give a distance. This can't happen with a prism because the EDM will only measure to the prism. An example may make the problem clearer. Suppose it is necessary to measure to the outside corner of a building 200 feet away. There is a fence 3 feet behind the building. We want to measure to the exact corner of the building. The building has vertical corner boards which give it a very sharp corner. If the instrument person is not very careful sighting the building corner and the cross hairs of the total station are slightly outside the corner, the EDM might measure the distance to the fence, not the building. The error would probably not be obvious to the instrument person because they would be unlikely to visually tell the difference between 200 feet and 203 feet. If the building was 100 feet away and the fence was 200 feet away the error would be obvious, providing that the instrument person actually noticed the distance measurement. This is an easy omission to make when using a data collector because the instrument person doesn't need to write a distance into a field book so the error is likely to go unnoticed. A simple way to prevent this from happening is to make a test shot on the building an inch or two inside the corner and make a mental note of the measured distance. Then, carefully move the crosshairs to the corner and take a second shot. If the distance doesn't change very much, the instrument person will know that the location is OK. Alternatively, if the accuracy of the building location is not that important, a point well inside the corner can be located instead of the actual corner.

Because it is not necessary to walk to each location with a prism, reflectorless EDMs can save a great deal of time when it is necessary to make many locations. Reflectorless total stations use more battery power than standard EDMs so total stations with reflectorless technology have the capability to switch back and forth between the two modes in order to extend battery life.

5.4 Robotic Total Stations

A robotic total station is a total station where the horizontal and vertical axes are driven by servo motors. They also incorporate communication devices allowing them to be controlled remotely. The robotic total station does not require that a person be at the instrument. Instrument controls are located on a data collector carried by the person who walks around locating physical evidence. The total station locks onto the prism and tracks it as a person moves from one point to the next. The robot uses its servo motors to turn the telescope. When the roving person finds evidence that he or she wants to locate, the prism is placed on the object. The person then presses a button on the data collector which instructs the robot to measure the angle and distance and transmit the data to the data collector where the information is stored electronically.

Robotic Total Stations have the potential to be economic in that they theoretically allow one person to perform a survey instead of the normal two to four person crew. However, from a practical point of view, carrying around all the equipment that surveyors need often takes at least two people anyway. The robotic total station can be much faster and efficient than a standard total station even if there is a 2 or 3 person crew.

One disadvantage of the robotic total station becomes apparent when the prism goes behind a tree or house. The total station will lose its lock on the prism. The roving person must then initiate a search routine and hope that the total station can find the prism again. If it doesn't, it may be necessary to walk back to the total station in order to regain lock. Some total stations are better than others in maintaining lock. Of course, this is not an issue if there is a person standing at or near the total station.

5.5 Servo Driven Total Stations

Somewhere between a robotic total station and a conventional total station lies the **Tracking Total Station**. It is similar to the robot except it does not use a radio to communicate data to a remote data collector. Instead, the data collector is located at the total station. Tracking total stations have servo motors just like the robotic total station, which allows the total station to automatically track the prism. The instrument person must remain at the total station, while a second person walks around making the locations. The advantage of a tracking total station is that the telescope tracks the prism as it moves from one point to the next. This makes locations very fast because the person at the total station does not have to aim and focus the telescope for each measurement. It is very useful for locating features where it is necessary to take many shots such as the edge of pavement and fences. For these locations, a person having a prism mounted on a prism pole simply walks from one point to the next and signals the instrument person when to take the shot. Measurements can proceed almost as fast as a person can walk. Losing lock is not a problem because the person at the total station can simply aim the total station at a point near the prism and it will automatically lock onto the prism again. Tracking total stations are also useful for finding a prism in dense undergrowth. There are times when the total station will lock onto the prism when it is barely visible when looking through the telescope.

Many servo driven total stations do not have traditional horizontal and vertical clamps to lock the telescope onto the target, only horizontal and vertical tangent screws. The instrument person can simply move the telescope by hand in the direction of the target. Once the telescope is pointed near the target, the tangent screws can be used to fine tune the direction. The principle is similar to a friction clutch in that turning the total station body and telescope by hand will overcome the tangent screw clamping force. Some total stations have two speed tangent screws which alternately allow course and fine adjustment.

Unlike transits and theodolites, all total stations require batteries in order for the electronics to function. Some total stations have onboard batteries that fit into the instrument. For other total stations, external batteries are required. Standard total stations do not require much power to operate so they often use small onboard batteries. Robotic total stations which have servo motors use quite a lot of power so the batteries must have much greater capacity. Often, these batteries are externally mounted and connected to the total station through a cable. Some total stations have battery packs which can hold several batteries at one time. This is particularly useful when a servo driven total station must turn many angles and measure many distances during a day. Most of the batteries in use are rechargeable. Some of the chargers operate from a conventional 115 V AC outlet and others from a 12 V vehicle DC system. The DC chargers are most convenient as they allow batteries to be charged in the survey vehicle in the field. Survey crews may have multiple batteries so that one or more will always be charging in the vehicle while the crew is using the total station. Some chargers and/or batteries may have temperature limitations for charging so extremely cold or hot weather may prevent charging in the vehicle.

5.6 Tripods and Tribrachs

Tripods are used to support and position surveying instruments and prisms accurately over a point. Tripods are also used to support GPS antennas. A tripod consists of three adjustable legs which allow it to be set up on rough and sloping terrain. A tripod can be seen in Fig. 5.4. The lower sections of the legs slide in and out of the upper sections. The upper sections have clamps which can be used to hold the legs at the desired length. Although it takes a few minutes to set up a tripod, it is very steady and rigid, so accurate and precise measurements can be made. When a total station or prism is set up over a point, it is crucial that it is perfectly level. For an instrument, horizontal angles must be measured along a plane that is perfectly horizontal. The most efficient way to level an instrument is to have the instrument attach to a tripod fixture which can be leveled. This fixture is called a **Tribrach**. The top plate of the tripod is a machined flat surface to which a tribrach can be attached.

A tribrach can be seen in Fig. 5.5. Tribrach dimensions are standardized so the instruments and prisms from different manufacturers can be attached to the tribrach interchangeably. There is a screw under the top of the tripod, shown in Fig. 5.6, which attaches the tribrach to the tripod. By loosening this screw, it is possible to slide the tribrach an inch or so in any direction over the top of the tripod. The long slot allowing adjustment can be seen just above the large black attachment screw. This adjustment allows the tribrach to be precisely centered over the point. The adjustability means that the tripod itself need not be perfectly centered over the point.

The tribrach is leveled using the three leveling screws until the bullseye level is centered. The bullseye level can be seen in Fig. 5.5. There is a small optical telescope on the tribrach (left side of the image) called an **Optical Plummet**. The optical plummet contains a magnifying lens with cross hairs and a right angle prism. The right angle prism is located exactly in the center of the tribrach so when the surveyor looks through the optical plummet she can see the point on the ground.

Fig. 5.5 A tribrach used to attach total station or prism to tripod

Fig. 5.6 Tribrach attachment screw. Hook for plumb bob string can be seen hanging below attachment screw

The optical plummet is used to visually set the tribrach exactly over the point. Some newer tribrachs have laser plummets which project a dot on the ground, thereby eliminating the need to look through an optical plummet. The disadvantage of the laser plummet is that it is necessary to replace the battery when it becomes discharged. With older instruments, such as transits, that did not use a tribrach, it was necessary to hang a plumb bob from the center of the instrument in order to center the instrument over the point. On windy days the wind would blow the plumb bob around, making it difficult to accurately position the instrument over the point.

Note that the optical plummet and bullseye level are attached to the top plate of the tribrach, not the bottom plate. The top and bottom plates are connected together with three leveling screws. This allows the top plate to be leveled even though the top of the tripod and the bottom plate of the tribrach are not level. The crosshairs in the optical plummet will only be centered accurately over the point when the tribrach is perfectly level. In other words, until the top of the tribrach is leveled, the line of sight of the optical plummet will not be vertical. It is therefore necessary to level the tribrach before the optical plummet can be relied upon to center the tribrach over the point. The tribrach leveling screws have a limited range, so when learning to set up a tripod, one of the more difficult lessons is to keep the top plate of the tripod as level as possible at all times. Of course, the tripod must also be kept over the point at the same time. Accomplishing this on a steep hill or on shaky ground in a swamp can be a challenge for the inexperienced. Figure 5.7 is an example of a difficult set up. The tripod legs are fully extended and the legs had to be set on stones on order for the tripod to be high enough to clear the monument.

Tribrachs are designed so that the instrument, prisms and GPS antennas can be affixed interchangeably. The device is placed on the tribrach and the locking lever (on the right side of Fig. 5.5) turned so that the device is locked to the tribrach. To exchange an instrument with a prism, the lever is unlocked, the old device removed and the new one inserted and the lever is re-locked. Because the equipment is precisely machined, various devices can be interchanged with no loss of accuracy. This interchangeability means that a tripod only needs to be set up once over a point. When traversing, a prism or total station can be interchanged using the same tripod. Not having to set up a new tripod saves a great deal of time, particularly when ground conditions make setting up difficult.

Tribrachs are designed so that the instrument, prisms and GPS antennas can be affixed interchangeably.

Tribrachs are not designed to handle abuse so, at the end of the day they should be removed from the tripod and stored in a safe place. To avoid scratching or denting the top plate of the tripod during transportation, it should always be kept covered with the protective cover (which should have been supplied when the tripod was purchased). The tripod top plate is usually machined aluminum so it is relatively soft and will scratch or gouge easily, potentially destroying the accuracy and usefulness of the tripod. Before installing or removing a tribrach from a tripod, always center the three adjustment screws so that the top and bottom plates are parallel. Some tribrachs have marks on the leveling screws which indicate the center position. Doing this insures that the next time a tribrach is used, full adjustment will be available.

The lower ends of most tripod legs have replaceable steel inserts which are sharpened to a point. This helps the legs to be pressed into the soil or hard pavement. Tripod legs also have protrusions just above the points which allow a person to stand on the leg in order to embed the points firmly into the soil. When setting

Fig. 5.7 GPS receiver set up over an ancient stone monument. Notice that the tripod legs are fully extended and that it was necessary to set the legs on stones in order to raise the tripod high enough above the monument. Multipath proved to be a problem with this setup

up a tripod, the instrument person will center it over the point, and ensuring the top is approximately level, use the full weight of the body on each leg to embed the point into the ground. If a tripod has to be set up on pavement, such as blacktop, it should be checked frequently for level. On a warm day, the sun will heat the blacktop making it soft, allowing the legs to sink in. Of course, this causes the tribrach to go off level and to move off the point. Another problem occurs when a tripod is set up on frozen ground where the legs will not penetrate the soil very much, if at all. If the sun comes out and thaws the ground surface, the weight of the tripod will cause one or more of the legs to sink in, moving the tripod off the point.

Caution is advised when setting up a tripod on a smooth surface, such as a concrete sidewalk or concrete floor. If the points on the end of the legs are unable to penetrate the surface, there is nothing to prevent one or more of the legs from skidding out, causing the instrument to come crashing down onto the ground. The steel points also become dull after use so they are less capable of gripping the surface and more likely to slip. Frequently replacing or sharpening the points is a good idea. On some tripods, the leg hinge pins at the top are adjustable to increase the friction, but this should not be relied upon to prevent the legs from skidding on a smooth surface. Having the hinges too tight makes setting up the tripod more difficult. One option is to drill small holes into the concrete for each of the leg points. Some tripods come with adjustable straps which attach at the bottom of each leg so that the tree legs can be tied together, preventing them from spreading.

5.7 Setting-Up a Tripod

Knowing how to set up a tripod is one of the first things a new member of a survey crew must learn. If the survey will be made using a total station and prisms, tripods will be used to support them. If the survey is to be made using static GPS, a tripod will be used to support the GPS antennas. If a GPS base station is to be used, a tripod will support the base station. Even small surveys may require many setups so the persons setting up the tripods will need to be both accurate and efficient. Where many setups are required during the course of a survey, the amount of time expended on setting up tripods can add up to many hours. Learning how to quickly and accurately set up a tripod will save the survey crew time and the client money. The following paragraphs are intended primarily for those who are learning to become surveyors or those who wish to know the details of setting up a tripod. Readers who do not plan to ever set up a tripod may want to skip this section.

Knowing how to set up a tripod is one of the first things a new member of a survey crew must learn.

Setting up a tripod is not difficult, at least on relatively level terrain, if the correct procedure is followed. Experienced surveyors each have their preferred method but there are common steps which all surveyors tend to follow. Once a person has performed many setups, they will develop a technique that works best for them. For this explanation, we will assume that the ground is relatively level. We will also assume that a tribrach has already been installed on the tripod. Because a total station is a very delicate piece of equipment it should remain in its case until the tripod setup is complete. *Never attempt to set up a tripod with an instrument in the tribrach.* Often, when the legs are stepped on in order to force them into the ground, they will glance off a stone creating enough shock and vibration to damage an instrument or throw it out of calibration. Although some survey crews have been known to leave a total station on the tripod while setting up, there is really no justifiable reason to do so, as it only takes 5 or 10 seconds to move the instrument from its carrying case to the tribrach. Setting up a tripod on level ground should only take a few minutes, and an expert can do it in a minute or two. If setups are averaging longer than 4 or 5 min, there is probably something wrong with the procedure being used.

When a tripod is moved from one setup to the next, or when using the tripod for the first time on the job site, it is customary to turn all three tribrach leveling screws to their center position so that the top and bottom plates of the tribrach are parallel. The tribrach attachment screw should also be loosened and the tribrach centered on the tripod top plate. This procedure only takes a few seconds and it insures that when the next setup is started, there will be full adjustment available. Tribrach screws should also be centered at the end of the work day when the tribrach is removed from the tripod and the equipment put away in the survey vehicle.

The basic steps in setting up a tripod are as follows:

1. Adjust the legs to the correct length so that the tripod will be at the proper height;
2. Roughly position the tripod over the point with the legs at the correct spread and height;
3. Move the tripod and/or swing the legs in or out so the tribrach is roughly level and approximately over the point;
4. Step on the legs to firmly embed them into the ground. This will probably move the tribrach off level and away from the point;
5. Re-adjust the leg lengths, if necessary, to bring the tribrach back to level and over the point;
6. Adjust the tribrach leveling screws to precisely level tribrach and,
7. Slide the tribrach over the tripod top plate until it is perfectly centered over the point.

Setting up a tripod is an iterative process. Each adjustment will affect another adjustment so it is necessary to repeat the adjustments incrementally until the desired result has been obtained. The adjustments are initially course, progressing to very fine and precise adjustments as the tribrach becomes level and centered over the point.

The first step is to start off with all three legs extended to the correct length. The easiest way to insure that all three legs are the same length is to keep the three legs together when adjusting their lengths. A simple method is to use one hand to hold the tripod off the ground at the desired height, loosen one leg clamp and carefully let the leg slide down to the ground. Don't just let it drop as this could damage it or dull the point. Tighten the clamp and allow the one leg to support the tripod. The remaining two clamps can then be loosened and the legs allowed to extend to the ground. All three legs will now have the same extension length. Another way to extend the legs is to start with the collapsed tripod standing on the ground, loosen all three clamps and lift the top section to the correct height. With the clamps loose, gravity will cause the lower sections to stay on the ground while the top section is lifted. All three clamps can then be tightened.

Remember that once the legs are spread apart, the tribrach will be lower than when the legs are together so this must be taken into account when extending the legs. If the tripod is set up on soft soil, stepping on the legs will cause them to sink in several inches so this must also be taken into consideration. For beginners, a good place to start is to have the top of the tripod at about chin height when the legs are together. The goal is to end up with the total station telescope roughly at eye level so the target can be comfortably sighted. Consideration must also be given to the types of measurements that will be taken which might affect the optimum height. For example, if it will be necessary to sight a point on the ground just a few feet away from the instrument, the telescope eyepiece may end up being quite high necessitating that the instrument person stand on their toes or on a box in order to see through the telescope. If there is a hill between the instrument and the target, it may be necessary to set the instrument higher than normal in order to see over the crest of the hill. If more than one person will be using the instrument and one person has a tall stature while the other is short, the instrument will need to be at a height which will accommodate both persons.

If the tripod is to be set up on the side of a hill, two legs should be downhill and one leg uphill. When measuring angles, it is easier to stand between two of the legs rather than having to straddle a leg. When many locations must be made in many different directions, straddling a leg may be unavoidable but if only one or two angles need to be measured, the leg orientations can be taken into consideration. When it will be necessary to use a tape to measure from the point over which the tripod is set, you must be careful to insure that one of the legs is not on the line to be taped; otherwise it will be necessary to move the tripod in order to make the measurement. The basic premise here is to simply think ahead and try to envision what will need to be accomplished during the setup and position the tripod to the best advantage.

Once the tripod legs have been extended, the tripod can be placed over the point. The legs are then spread apart similar to the tripod shown in Fig. 5.8. If the legs are too close together, the tripod will not be as steady as if the legs have a substantial spread. Once the legs are spread apart the tripod can be moved over the point as follows. Stand between two of the legs and grab the third leg, which is now the furthest leg from you, with one hand. Move in close to the tripod so the two closest legs are pressing against the body. The method used in lifting a tripod is shown in Fig. 5.8. This procedure should be practiced until the tripod can be lifted off the ground without changing the leg spread. Once lifted off the ground, the tripod can be set over the point. Positioning the tripod over the point is usually done visually, however a plum bob can also be used. Some tripods have a hook for attaching a plum bob string as shown in Fig. 5.6. Using a plumb bob takes a little longer but it allows the tripod to be easily located over the point, rather than visually guessing at the location. If a plumb bob is used, it is customary to tie the string using a slip knot so that the plumb bob can be raised and lowered easily. For most easy setups it is faster to position the tripod by eye than to use a plumb bob. For difficult setups, however, a plumb bob or laser plummet can be handy. With the tripod over the point, the top plate of the tribrach is now leveled by lifting each leg slightly off the ground and pivoting it in or out so that the bullseye level in the

Fig. 5.8 Lifting tripod. Notice that top of tripod is kept level

tribrach is nearly centered. Because the bubble is extremely sensitive, it will not be possible to perfectly center the bubble simply by moving the legs. For now, getting it somewhere near center is good enough.

Next, look through the optical plummet. If the point cannot be seen, place your foot under the tripod and move it around until the toe can be seen through the optical plummet. This will tell you how far away the point is from being directly under the tribrach. Lift the tripod again, being careful not to let the legs collapse, and move it toward the point. If the tripod started off relatively level and the legs were not allowed to move when it was lifted, it should still be roughly level—assuming the ground is level. If not, pivot one or more of the legs again so that the bubble is centered. Look through the optical plummet again. Repeat this procedure until the optical plummet shows that the tripod is very close to being over the point—preferably within a few hundredths or so. Remember that the tribrach must be approximately level before the optical plummet can be used. At this stage it is only necessary to have the tripod close to the point and roughly level. Do not waste time trying to get it perfect because once the legs are stepped on it will move.

Once the tripod is nearly over the point, it is necessary to step on the legs so that they penetrate the ground sufficiently to provide a rigid setup. This is done by using the full weight of the body. Stand on each leg until it sinks in to the ground. You may need to jump on each leg a little bit if the ground is hard. If the density of the ground is different under each leg or there are stones in the ground, each of the legs will penetrate to different depths. This will cause the tripod to go off level and move away from the point. This is where many novices run into trouble. It is important to recognize that, *if the tripod was level and over the point before the legs were stepped on, it can be brought back to that position simply by readjusting the leg lengths*. When the tripod goes off level when stepping on the legs, do not pull the legs out of the ground and start again. This would be a waste of time. Once you are confident that the legs have been fully seated in the ground, watch the bullseye level in the tribrach and adjust the leg lengths, one at a time, to bring the bubble back to center. Do not use the leveling screws at this point to try to level the tribrach. If the tripod was level and over the point before the legs were stepped on, adjusting the leg lengths to level the tribrach will put it back over the point.

The tribrach should now be roughly level and very close to the point. Now you can use the tribrach leveling screws to perfectly level the tribrach. Keep in mind that the bullseye bubble will move in the same direction as your left thumb when the leveling screws are turned. If the setup was carefully performed up to this point it should not be necessary to move the leveling screws very much. If you find that you run out of thread on the leveling screws, you are probably not being careful enough ensuring that the tripod top plate was level when the legs were first placed over the point. Another possibility is that the tripod screws were not returned to their central position before the setup was started. There are occasionally very difficult setups where it is not possible to get the tripod top plate as level as it should be and the tribrach leveling screws are near the end of their travel. If this is happening to you most of the time, however, you are probably not following correct procedures.

Once the tribrach is level, loosen the tribrach attachment screw and, while looking through the optical plummet, slide it over the tripod top until the optical plummet shows that the tribrach is exactly over the point. When sliding the tribrach, be careful not to rotate it as this will cause it to go off level, particularly if the top of the tripod is considerably off level, i.e., the tribrach top and bottom plates are not close to being parallel. When the tribrach is over the point, tighten the attachment screw and re-check the bullseye level and optical plummet. It may be necessary to repeat this fine tuning a few times until the tribrach is perfectly level and over the point. When finished, **always** check to be sure that the leg adjustment screws are tight and the tribrach attachment screw is tight before putting the instrument on the tripod. Needless to say, it would be exceedingly embarrassing, and possibly very expensive, to have a perfect setup spoiled when the weight of the total station causes one of the legs to collapse because you forgot to tighten a leg screw.

The reader must be aware that the bullseye level on the tribrach is not as accurate as the level on a total station or theodolite. Although the bubble is accurate enough to be used for centering the tribrach over a point, the instrument level must always be used for leveling the instrument. This means that each time an instrument is set on a tribrach, the instrument person must fine tune the tribrach by adjusting the tribrach leveling screws. For some instruments, the level is so precise that the slightest amount of turn on a tribrach screw is enough to cause the instrument to go off level. If you are having trouble centering the instrument bubble, one method is to turn the screw a tiny bit then remove the hand from the screw and wait a few seconds for the bubble to settle down. Sometimes, the weight of the hand or pressure from the fingers is enough to disturb the level. When leveling an instrument it is necessary to turn the instrument 180°, check the bubble a second time and, if it is off to split the difference. If, after leveling the instrument, the tribrach level is seen to be off, the tribrach bubble will need to be adjusted.

Total stations and theodolites have two levels positioned 90° apart. One of the levels is parallel to the telescope and the other level is 90° to the telescope. The best way to level the instrument bubble is to align the level bubble parallel to two of the tribrach leveling screws. One of the two screws can then be turned to center the bubble. The third screw will then be turned to center the other bubble. This method allows each level to be adjusted without affecting the other level. If you find that, when you are adjusting one bubble and the second bubble moves, then the level is not quite parallel to the two tribrach screws.

5.8 Corner Cube Prisms

Prisms are used to reflect light emitted from the total station EDM back to the total station so that the distance between the total station and the prism can be measured. Figure 5.9 shows a prism on a tripod. The prisms used for surveying are **Corner Cube Prisms**. A unique feature of corner cube prisms is that even if the prism is not perfectly aligned with the light source (EDM) it will reflect the light beam directly back to the source. All prisms have a **Prism Offset**, sometimes called a **Prism Constant**. The prism offset is a distance correction that must be applied to all slope distances measured with a particular prism.

Light from the EDM has to travel on a path into and out of the prism in order to be reflected back to the total station and this affects the prism offset. This can be seen in Fig. 5.10. The speed of light through the air and the glass of a prism is different so this is an additional correction that must be made. The mounting point of a prism in relation to the plumb line of the prism holder also affects the prism offset so this must be taken into consideration when calculating the distance between the prism and total station. This is shown in Fig. 5.10 as the vertical axis. Prism manufacturers provide prisms with different offsets. Some prism holders allow the offset to be changed. Normally, the prism offset is programmed into the total station or data collector software so that distances are automatically adjusted for the offset.

Fig. 5.9 Prism that reflects light back to total station. Prisms can be mounted on tripods or on a portable prism pole

5.9 Prism Poles

When the point to be measured to is a traverse point, the prism will be mounted on a tripod, because a tripod is very stable and, as we have seen, it is possible to set a tripod accurately over a point. If the thing to be located is a fence, driveway, hedge row or other feature where accuracy is not so demanding, the prism can be mounted on a **Prism Pole**. A prism pole can be seen in Fig. 5.11. Prism poles have built-in bullseye levels which allow the pole to be held vertical so an accurate location can be made. However, prism poles do not have to be "set up" like

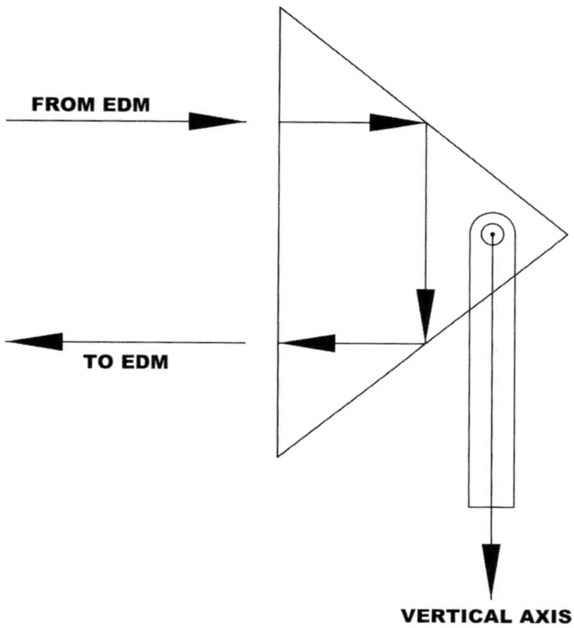

Fig. 5.10 Prism offset

a tripod. A person simply walks along and places the pole on the feature to be located. The person on the total station then makes the measurement and records the location in the data collector. Surveyors often call this **Taking a Shot**. Because the prism pole cannot be held perfectly steady it is not considered to be as precise as a tripod mounted prism, but it is much faster, lighter and easier to carry. Notice in Fig. 5.11 that a circular array of small prisms allows a measurement to be taken from any direction. The person holding the pole need not be concerned whether or not the prism is aimed exactly at the total station. Not all prism assemblies have multiple prisms. Sometimes a single prism is mounted on a prism pole. In such cases the person holding the pole must be careful to aim the prism at the total station. Bipods are sometimes attached to prism poles. A bipod has two adjustable legs and a clamp which allows it to be attached to any prism pole. Similar to a tripod, the bipod legs can be pressed into the ground to provide a rigid support for the prism pole. Setting up a prism pole and bipod is slower than taking a shot with just the pole, but it is faster than setting up a conventional tripod. Bipods also weigh a lot less than tripods so they are easier to carry. A bipod can also be quickly removed from the prism pole when it is no longer required.

Because the prism pole cannot be held perfectly steady it is not considered to be as precise as a tripod mounted prism, but it is much faster, lighter and easier to carry.

Fig. 5.11 Prism pole. The clear cylinder with the gold reflector contains a circular array of LEDs which provide a light source for tracking. A *circular array* of small prisms can be seen a few inches above. The *circular bullseye* level is at the *bottom*. The black knob allows the height of the prism to be adjusted

Surveyors are aware that prism poles are subject to quite a lot of abuse in the field and the bullseye levels are in constant need of adjustment. If the locations to be made need to be accurate, it is always a good idea to check the prism pole to be sure that the bullseye level is adjusted properly before the pole is used. This should be done at least once every day. The adjustment is easily accomplished by using a bipod and setting up the prism pole over a point. The vertical cross hairs of the total station can then be used to sight the edge of the pole. If it is not exactly vertical, the bubble is adjusted until it is. The pole is then turned 90° and the procedure repeated, adjusting the bullseye level as necessary. The process is repeated until the pole is vertical in both directions.

5.10 GNSS and GPS

The **Global Navigation Satellite Systems (GNSS)** is a constellation of orbiting satellites and ground stations which monitor and communicate with the satellites. GNSS is used for navigation and positioning on the earth's surface. Presently, GNSS consists of satellites owned by the United States and Russia. The Global Positioning System or **GPS** is the United States portion of the system. The Russian Global Navigation Satellite System or **Glonass** is the Russian portion of the system. Satellite systems from other countries may be added to GNSS in the near future. The advantage of GNSS over GPS is the availability of additional of satellites which can make positioning more accurate. It can also shorten the amount of time required for observations. Some older GPS receivers only work with the GPS system, however many newer receivers work with GNSS. In this book, the terms GNSS and GPS are used interchangeably to mean positioning by means of satellites.

The GPS receivers used by surveyors receive signals from the satellites orbiting the earth. The satellites continually transmit a precisely timed message which includes the exact time the message was transmitted and the exact satellite position at the time that the message was transmitted. The receivers are able to use this information to calculate the distances from the receiver to the satellites. If enough satellites are available and if they are spread out over the sky, the receiver is able to use the distances from each of the satellites to calculate an intersection representing the receiver's location on the surface of the earth. Although survey grade satellite receivers were initially very expensive, the cost has decreased markedly over the years and they are now so affordable that most boundary surveyors use them. Figure 5.12 shows a static GPS receiver on a tripod.

There are two common types of GPS surveys: Static GPS and Real Time Kinematic (RTK) GPS. Survey grade GPS should not be confused with the common hand held or vehicle mounted GPS receivers. The antenna quality and software used with survey grade receivers enables vastly improved accuracy. For example, while a hand held GPS receiver may provide a position within about 10 m (about 33 feet), survey grade receivers are capable of centimeter level accuracy.

Static GPS is the most accurate method and it is used for control networks and when precise positioning is required. The disadvantage of static GPS is that the receivers must occupy a point for a long time—from 30 min to several hours. Generally, the longer the receiver occupies a point, the greater the accuracy.

Fig. 5.12 Static GPS receiver capable of centimeter level accuracy

Occupation times increase as the distance between receivers increases. After the data is collected, it is uploaded into a personal computer for **Post-Processing**. During the post-processing phase, the software will make the necessary adjustments and corrections and calculate coordinates for the occupied points. Accuracy for static GPS[1] can be about one centimeter horizontal and approximately 2–4 cm vertical, however environmental factors may reduce this accuracy in practical

[1] Based on a 95 % confidence level. See Standards and Guidelines for Land Surveying using Global Positioning System Methods, August 6, 2004. The Survey Advisory Board for Washington State Department of Natural Resources.

applications. Greater accuracies may be claimed by GPS manufacturers or sellers, as the technology is constantly improving. Prior to acquiring a system a surveyor would probably want to get some hands-on experience with a particular system to see if it lived up to his expectations for the type of work in which the system would be used. As we discussed earlier, GPS accuracy is usually measured statistically, in terms of probabilities, not linearly as with conventional traversing. Comparison of the two surveying methods is possible when the distance between GPS points is taken into consideration. When GPS accuracy is equated with linear accuracy, it is common for longer baselines to achieve accuracies in excess of closure standards commonly accepted for boundary surveys. Recall that for conventional boundary surveys; commonly accepted closures are in the 1 in 10,000 to 1 in 15,000 range.

RTK GPS consists of two receivers. One receiver is set at a known point, for example a National Geodetic Control (**NGS**) point. This receiver is referred to as the **Reference Station**. The other receiver is referred to as the **Rover**. The rover is usually an antenna and receiver mounted on a prism pole. The reference station remains fixed over the reference point during the survey while the rover is moved around to locate evidence or physical features or to set points. Occupation times for the rover vary from 180 seconds to as short as 15 seconds. Expected accuracy (see Footnote 1) is 1 to 4 cm horizontal and 2 to 6 cm vertical.

Survey grade GPS receivers work by **Differential Positioning**. Differential GPS requires the use of multiple receivers. Atmospheric and other effects cause a degradation of the GPS signal which affects GPS positioning accuracy. However, if a number of receivers are located in relatively close proximity to each other (a few miles), each receiver will be subject to the same accuracy degradation. If one of the receivers is located at a known reference point, such as an NGS control point, which has known state plane coordinates and a known elevation, the GPS coordinates of the receiver on the control point can be compared with the known coordinates of the control point and correction factors can be calculated. The correction factors can then be transmitted to the rover so the GPS coordinates of the rover can be adjusted.

When using RTK, the correction factors must be continuously transmitted to the rover in order for the corrections to be applied in real time. This requires a radio link between the reference station and the rover. For static GPS receivers, the correction factors are applied during post-processing so no radio link is required. As GPS has become more commonly used, continuously operating reference stations (**CORS**) have become available in some locations. These stations are maintained by the National Geodetic Survey. If a GPS survey is made in the proximity of a CORS station, the data can be downloaded from the NGS web site for post-processing. In some locations, services are available so that the CORS correction data or privately owned correction data can be transmitted to the rover in real time using a cell phone or other cellular device with a data plan. Some newer GPS receivers include an integrated cellular connection so that an external cellular device is not required. In addition to the NGS CORS stations, private, university and state CORS stations may be available in certain locations. This eliminates or reduces the need for the surveyor to set up a receiver at a reference station when using RTK. RTK receivers can also work in static mode.

A number of factors affect GPS accuracy. Atmospheric effects such as interference in the ionosphere, and the amount of water vapor have an effect. Multipath can have a substantial effect on accuracy. Multipath is caused by GPS signals reflecting off of large structures such as flat sided buildings, signs and trees. Multipath causes multiple signals to be received—those from the satellites and those which bounce off the reflectors. These signals can confuse the software and cause errors in position. Even parking a survey vehicle next to a GPS receiver can cause multipath problems.

The number of satellites visible and the geometry of the satellite constellation is an important factor to consider when using GPS. For high accuracy, it is recommended that a minimum of 5 satellites be available. It is best if satellites are spread out across the sky rather than grouped into a small cluster in one section of the sky. Mission planning software is available which will tell the surveyor what time of the day or night will provide acceptable satellite geometry. As the satellite geometry is constantly changing, the time required for the observations will need to be factored in to ensure that a favorable geometry will be continuously available throughout the observation period. The Positional Dilution of Precision (**PDOP**) value is an indication of the impact of satellite geometry on GPS precision. The smaller the PDOP value, the better the satellite distribution, and the better the precision. For survey grade GPS, many consider that a PDOP of 4.0 or less is adequate.

For boundary surveys, multipath effects and attenuated signals caused by tree branches and leaves (canopy) can reduce GPS accuracy below that which is acceptable. Canopy may also reduce the number of available satellites which can be used. These factors will limit the usefulness of RTK GPS when working in or next to wooded areas. In such cases, the system may not deliver the accuracy necessary for boundary surveys. Static GPS will also be affected in the same way; however, because of the long occupation times, it is likely that static GPS will be used primarily for setting up control networks, not for setting lot corners in real time. This provides greater flexibility because sites in open areas can often be selected for the control points. This will help to avoid multipath problems.

Where obstructions will unacceptably reduce GPS accuracy, traditional surveying techniques using a total station will still need to be employed. However, as we will discuss further on, it is possible to combine GPS with conventional traverse techniques. GPS receivers are now so economical that they should be included in every boundary surveyor's kit. GPS will allow small or large boundary surveys to easily and economically be tied into the state plane coordinate system. This will provide additional redundancy should it become necessary to reestablish boundary corners at some later date.

5.11 Data Collection

In the old days of boundary surveying, it was necessary to record all observations into a **Field Book** by hand. Modern surveyors now use **Electronic Data Collectors**. A data collector is a small computer that records information sent to it by the total station. Typically the person running the total station points the telescope at the target and presses a button on the data collector. The data collector instructs the EDM to measure the distance. Once the distance is measured, the horizontal and vertical angles and the distance are transmitted to the data collector where the information is stored in memory. If the data collector is connected to a GPS receiver, the coordinates and elevation of the point are stored in memory. When the field crew returns to the office, the data is electronically transferred from the data collector to a desktop computer. Now, with the increased use of cell phones and data storage in the cloud, the field data could be made available to the office, which could be hundreds of miles away, in real time.

Data collectors normally record two basic types of data. The first data type consists of the **Raw Data** which is collected by the total station. For a total station this includes the horizontal angle, slope distance and vertical angle. The data collector will also store the name of the backsight, occupied point and foresight. This will be explained in more detail in the section on using a total station to measure an angle. Raw data is often stored in an ASCII text file which can be read by most word processors or a simple text editor on a personal computer.

In addition to the raw data file the data collector calculates and stores the 3 dimensional coordinates of the point, the name or number of the point and a description of the point. Coordinate files are often proprietary binary files which usually cannot be read with a word processor. The purpose of keeping a raw data file is that the file is a record of every measurement which is taken in the field. Even if the coordinate file were to become lost or corrupted it would be possible to regenerate it using the raw data file.

5.12 Field Books

The information in the raw data file could also be written in a field book. Before data collectors became available, field books were the commonly accepted way of recording data. When field books were used, mistakes were sometimes made reading the angle or distance from the instrument, transposing numbers or writing incorrect information into the book. Because the data is transferred automatically, data collectors have largely eliminated these errors. On the down side, it is not unheard of for data collectors or inexperienced users, to mess up and lose data that may have taken hours or days to collect. This can't happen with a field book, although it is possible that someone could lose it. For certain surveys where evidence must be admissible into court it may still be preferable, or even essential, to have a written record of the field work, so, under these circumstances, a field book would still be necessary.

As already noted, before data collectors became available, field books were the only way to record measurements. Field books are usually hard cover, bound books which are small enough ($6'' \times 8''$) to fit in a large pocket. The pages are usually bound in pairs with one side having lined columns and a facing page with graph paper squares, as shown in Fig. 5.13. The graph paper is used for drawing

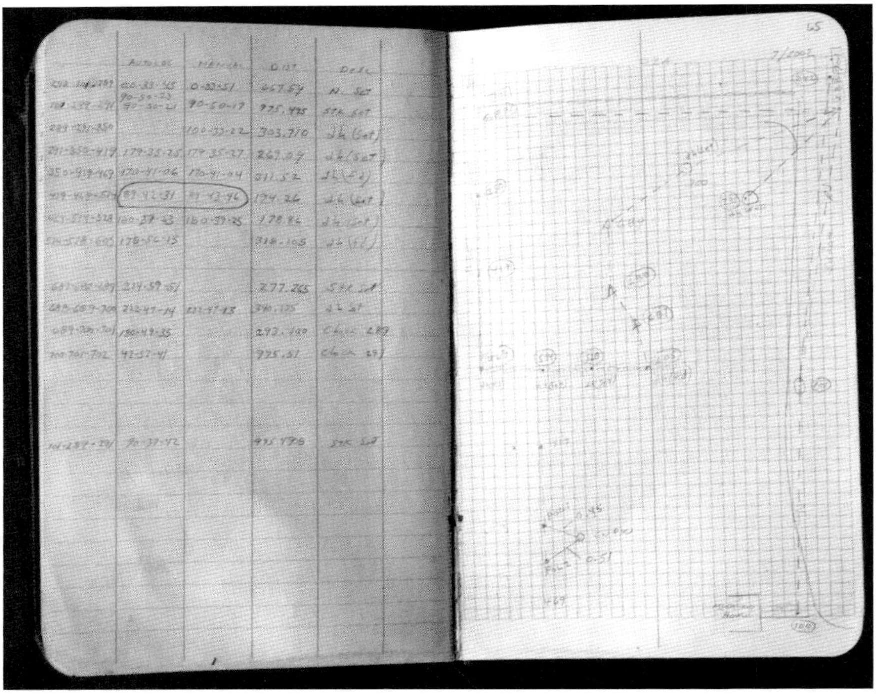

Fig. 5.13 Bound field book

images of the survey lines and evidence collected while the columns are used to record angles, distances, point names and descriptions of evidence. Field books are available with different types of rulings (columns and squares). The books usually contain water repellent paper so they can be used in inclement weather. Because the books are sewn and bound, the pages form a permanent record. Many field books contain 60 or 80 pages, so records for many surveys can be kept in a single book. Sometimes loose leaf field books were used. In this case the pages for a particular survey were often removed and kept in the folder with the remainder of the survey records. Surveyors must always create a table of contents in each bound field book showing the page numbers for each of the surveys. When a survey office has been in business for many years and it has accumulated a large number of field books, it is necessary to have an index of the surveys and the field book number and page corresponding to the survey.

Field books serve as important and unique records of boundary surveys. For many surveys, plans are recorded in a public recording system for the purpose of memorializing the results of a boundary survey. However, plans often do not show all of the information and evidence that was gathered during the field work. Traditionally, this information was written into a field book. Where a public agency, such as a local highway department, performed field work, the field book would sometimes be stored in a vault or other fireproof and secure location. These field books are often indexed and made available to both public and private land surveyors as evidence of the surveys that were performed by the original surveyor. When performing surveys of the public lands, field books of the original surveys are an integral part of these surveys and the surveyor must obtain copies of the relevant field books when performing a survey.

Many private land survey offices have collections of field books going back many years. Because these field books represent valuable and irreplaceable evidence of old surveys, the collections are often purchased from a retired surveyor by other land surveyors who may be called upon to perform retracement surveys of the original parcels, or nearby properties. Field books also contain evidence of boundary marker locations and ties to monuments which may help in recovering these monuments many years later. Sometimes, a retiring surveyor or his estate will donate or sell field books to a recorder's offices or a historical society for safekeeping and to be made available to current and future surveyors.

Digital data collection poses a potential problem for future boundary surveyors, and the public, because the records are, in some ways, less permanent and more likely to be lost than field books. The data files may be in a format only readable by certain software programs which may not be available twenty years or more later. Files may be stored on Compact Disks (CD) or Digital Video Disks (DVD). CDs and DVDs have uncertain life expectancies, which are dependent on factors such as type, manufacturing quality, quality of recording, handling methods and environmental conditions. Life estimates vary between 25 to 200 years. Unfortunately, there is often no easy way to tell if a particular disk has deteriorated because error detection capability built into the system will correct a certain number of errors until the point is reached when the disk is no longer readable.

Digital data collection poses a potential problem for future boundary surveyors, and the public, because the records are, in some ways, less permanent and more likely to be lost than field books.

Even though it is time consuming, surveyors using data collection should try to make it a habit of keeping written or printed records of all important field work. If no permanent written and bound written records of field work are kept, there is a likelihood that important records will disappear and not be available to future generations of boundary surveyors.

5.13 Some Common Terms Relating to Using a Surveying Instrument

Before we discuss how angles and distances are measured we need to learn a few terms used by surveyors which relate to the use of an instrument. The **Instrument** is the total station, formerly a transit or theodolite. The instrument is also called a

Gun, because it takes **Shots** on **Targets**. A **Target** can be a prism, a prism pole or a crew member holding a plum bob string. The target is the thing that is sighted by the instrument person for the purpose of measuring angles and, when the target contains a prism, for measuring distances.

Line is the direction in which the instrument telescope is pointing. More precisely, it is the direction in which the vertical cross hair is pointing. When a target is exactly centered in the vertical cross hair, the target is said to be **On Line**. **Taking Line** means sighting a fixed target with the instrument. Taking line is accomplished by moving the horizontal tangent screw on the instrument until the vertical cross hair on the instrument is perfectly aligned with the fixed target.

The meaning of the term **Give-Me-Line** depends on who is saying it. Give-me-line can be a request by the instrument person to the rod person to hold the target over a point so the instrument person can sight the target and set zero on the instrument. **Setting Zero** means locking an angle of zero on the instrument vernier or display when the telescope is pointed at the target. It is also referred to as **zeroing** the instrument or saying the instrument is **zeroed**. For example, when measuring an angle with the instrument it is first necessary to zero the instrument on the target from which the angle is to be measured. **Give-me-line** can also be a request by the instrument person to the rod person to hold the target over a point so the instrument person can measure an angle (and usually a distance) to the point. So, when the instrument person asks the rod person to give-him-line on a particular stone bound, the rod person must walk over to the stone bound and hold the prism pole on the bound so that the instrument person can locate it. To **Locate** means to turn an angle and distance to some object, usually some piece of physical evidence such as a stone monument or drill hole.

Give-me-line can also be a request by the rod person to be put on line by the instrument person. In this case, the direction of the line has already been fixed by the instrument person. As an example, if we need to establish a point between a distant target and the instrument, the instrument is first locked onto a target. The rod person then stands between the instrument and the target. The instrument person instructs the rod person to move right or left until the rod is exactly on line. Another example would be where the initial field work and calculations have been completed and the crew has returned into the field to set the final lot corners. The instrument person sets the instrument on a traverse point and calculates the angle and distance from the traverse point to the lot corner to be set. The instrument person then turns the instrument to the correct angle and instructs the rod person to move to the line so she can set the point. In this case, in addition to being on line, the rod person will also need to be a specific distance from the instrument, so the EDM will be used to measure the distance from the total station to the prism pole.

When the instrument person gives line he must communicate with the person holding the target in order to tell her to move left or right so that she is on line. Some surveyors prefer to tell the rod person to move north or south or east or west rather than left or right because left or right depends on whether the rod person is facing the instrument or whether she has her back to the instrument. Additionally, when using left and right, the direction is always relative to the rod person not the instrument person, so when the rod person is facing the instrument, as she usually is, the instrument person must remember to tell her to move in a direction opposite to the instrument person's left or right. Of course, if north, south, east or west is used, both the instrument person and rod person must be aware of which way north is. This shouldn't be a problem though—they are, after all, land surveyors.

A **point-on-line** is a point that is placed exactly on an established line. The line could be a property line or a traverse line. A point-on-line is different than a traverse point because the angle running through the point-on-line is always 180°.

Setting Up the Instrument or just **Set Up** means to set the instrument over a point. As we have seen, this requires precisely leveling the instrument and tribrach and precisely positioning the instrument and tribrach over a point.

All angles require three points: a **Backsight**, an **Occupied Point** and a **Foresight**. The occupied point is the point on which the instrument is set up. The backsight is the starting point from which the angle is measured. It is the point where the instrument is zeroed. The foresight is the point being measured to or being set.

Turning an Angle means the instrument person uses the total station to measure or set an angle on the instrument. When **Measuring an Angle**, three points are fixed and the angle is measured between them. When **Setting an Angle** only two points are fixed and the third point is to be set. Details of how this is accomplished will be explained further on. Older instrument circles are numbered in both directions so angles can be read either left or right. When recording angles it is important to note which direction the angle was turned. In data collectors, stored angles are preceded by AR or AL for Angle Right or Angle Left. Total stations normally measure all angles to the right from 0 to 360°.

When **Taking a Shot**, the instrument is used to measure the angle and distance to a point, usually to locate evidence. Locating evidence is also called a **Location**, so the instrument person is sometimes said to be **Making Locations** or **Performing Locations**. Taking a shot is also called a **Sideshot**. Often, multiple sideshots are performed at a particular setup. Zero is set on a backsight and multiple angles and distances are measured to various forms of evidence. For example, we may need to locate a stone bound, a drill hole and the corners of a house. If all of these points can be seen from one setup, sideshots will be used to measure to each one. In such cases, it is only necessary to set zero on the backsight once. In contrast, when traversing, it is only necessary to measure a single angle to the foresight. The instrument will then be moved to the next point. In traversing it is common to measure multiple angles between the backsight and foresight in order to increase precision but there is still only one foresight, so the end result is a single (averaged) angle. This measurement is called a **Traverse Shot** to differentiate from a sideshot.

The distinction between a sideshot and a traverse shot is primarily important when using a data collector. When taking sideshots, data collector software will increment the point numbers of the foresight after each shot. However, it will not increment the numbers of the occupied point and backsight, because they do not change. When a traverse shot is taken, the instrument only turns a single (usually averaged) angle to the foresight. The instrument is then moved to the foresight and the previously occupied point becomes the backsight. So when traversing, the data collector increments the point numbers of the backsight, occupied point and foresight. When taking shots with a data collector, the user is given the choice whether to take a sideshot or a traverse shot.

5.14 Improving Accuracy When Measuring Angles

Before we discuss methods for measuring angles with a total station or theodolite, it is important to have an appreciation of just how precise surveying instruments are, and how little movement it takes to dilute this accuracy. We noted earlier that one of the great challenges in boundary surveying is making extremely precise measurements under difficult environmental conditions. The survey crew therefore needs to be aware of factors which can reduce measurement accuracy.

We already discussed the fact that a 1 second total station is able to resolve about 1/16 of an inch at a distance of 1,000 feet. With this in mind, it should be obvious that the slightest disturbance of the total station will cause the crosshairs to move off the target. When measuring angles it is crucial that the techniques being used do not reduce the accuracy of the measurements. Experienced instrument persons understand this and they are sensitive to the factors which have the potential to affect the accuracy of the angles being measured. It is important that other members of the crew also have this awareness, particularly when they are working in the vicinity of the instrument.

One of the most important factors which has the potential to affect the accuracy of an angle is the firmness of the ground upon which the instrument is placed. The softer the ground is, the more likely the instrument will move. Movement of the instrument can be caused by the weight of the instrument allowing it to slowly settle into the ground. In other cases, the ground may be stable and firm enough to support the instrument without its moving, but the weight of a person walking around the instrument will cause the ground to move enough so that the angle accuracy is affected. People weigh several times more than an instrument and tripod so there is great potential for body weight to affect measurements, particularly when set up on soft ground.

We have discussed the effects of thawing ground and softening pavement due to heating by the sun as conditions which can cause the tripod legs to settle into the ground. If the tripod were to move, ever so slightly, between sighting the backsight and foresight, the angle being measured will not be correct. Although not commonly seen these days, some survey crews have used tripod “shoes”. These were simply thick plywood or boards, 6 or 8 inches square, which had an indention for the tripod leg points. They could be nailed into the pavement to prevent them from slipping. The large surface area distributed the weight of the tripod and instrument over the pavement, reducing surface loading and tripod movement.

Setting up on soft ground is sometimes unavoidable. It may be necessary to traverse through a swamp for example. In such cases it is often possible to find small islands, or raised areas, formed by tree roots and vegetation upon which the instrument may be set. These areas may still be soft but they may be the best choice available. If the raised area is next to a living tree and it is windy, there is a good chance that the movement of the tree in the wind will be enough to affect the angle. The author recalls such a setup. When looking through the instrument telescope the crosshairs could be seen moving from one side of the target to the other in concert with the tree swaying in the wind. Although the upper branches of the tree could be seen waving in the wind, there was no perceptible movement of the large trunk. Nevertheless, the movement was enough to make it impossible to accurately measure the angle.

When set up on soft ground, the instrument will move when the instrument person walks around the tripod. If the angle to be measured is close to 180° the instrument person must walk around the tripod in order to sight the target. If the ground is soft, the change in pressure on the ground from the instrument person's body weight will move the instrument. Sometimes the angle is small enough that the instrument person can stand in an intermediate position where both targets can be sighted without moving. In some cases, merely shifting body weight from one leg to another will be enough to cause errors in the angle. Where the angle is near 180°, one solution is to have two people measure the angle. Each person takes a position on opposite sides of the tripod. One person can then sight the backsight. The second person will then sight the foresight. Each person must be careful not to move while the angle is being measured.

When set up on soft ground, the instrument will move when the instrument person walks around the tripod.

Even when the ground is relatively firm, some general rules apply to working with an instrument. The instrument person should never place any weight on the tripod. When sighting angles, the tripod should not be grasped with one hand in order to steady the body. When it is necessary to walk around the tripod in order to turn an angle, it is best to walk in a large circle around the instrument rather than stepping in the vicinity of the tripod legs. When measuring angles, keep feet away from tripod legs and try to place them between the legs. In fact, never step on the ground next to a tripod leg. Everyone on the survey crew, except for the person running the instrument, should stay a reasonable distance away from the instrument. There is rarely a need for a crew member, other than the instrument person to be standing near the instrument.

One example where there is an understandable tendency to touch the tripod is when the ground is soft. The instrument person may try to stand a little further away from the instrument in order to avoid disturbing the ground. She then must lean forward in order to look through the telescope. In order to keep balance, there is a temptation to grab one of the tripod legs. Doing so will almost certainly move the instrument.

On warm pavement or on thawing ground, movement of the instrument may be relatively slow. In such cases better accuracy can be achieved by measuring the angles as quickly as possible. Just before measuring any angle, the instrument level should be checked and the optical plummet checked to insure that the instrument is still over the point. Never assume that zero is still set on a backsight. Always check zero before measuring an important angle. In any situation where there is a possibility that the instrument might move, it will be necessary to turn multiple angles or sets. In this way, an inaccurate angle will be noticed immediately because the precision will be poor. Without multiple angles, a bad angle might not be noticed until the crew is back in the office and it is discovered that the traverse will not close. This usually means another trip back to the swamp.

Tracking total stations help solve some of the problems encountered on soft-ground setups because the instrument person doesn't need to look through the telescope. She can stand a short distance away while the total station automatically locks on to the prisms and measures the angle. Some total station manufacturers claim that the ability of their total station to lock onto a target is superior to that of the human eye. This claim may be true or not true, but I would not want to be the one sitting in a witness chair during a trial concerning an incorrectly located lot corner, trying to explain to a jury why I did not look through the telescope to see if the total station was actually pointing at the target.

If many locations must be made during a setup, it is wise to occasionally go back and check zero on the backsight, particularly if there is any reason to suspect that the instrument may have moved. It is disappointing to find that, when checking the backsight after 50 locations have been made, the backsight it is substantially off. In such cases there is usually no way to tell which locations are good and which are bad. In other words, it cannot be known exactly when the instrument moved. When this happens the only choice is to take all of the locations again. This sad state of affairs is easily avoided by frequently checking the backsight whenever an opportunity presents itself, for example when the rod person will need some time in walking to another place before resuming locations. Checking the backsight should only take a few seconds so frequently doing so will not interrupt the flow of the work very much. It should be standard practice for the instrument person to always check the backsight as a final step just before the instrument is taken off the tripod when the locations are complete. Whenever the instrument person believes that the instrument may have moved he or she must immediately report it to the party chief, who will then decide what action to take.

It should be standard practice for the instrument person to always check the backsight as a final step just before the instrument is taken off the tripod when the locations are complete.

5.15 Example of Using a Total Station to Measure an Angle and Distance

This example assumes that a data collector is being used to record the information. However, a field book could also be used. Referring to Fig. 5.14, the backsight is a prism set at Point 4. The occupied point is Point 1, where the total station is set up. The Foresight is Point 2 where another prism is set.

Let's look at the step by step procedure for measuring the angle and distance shown in the Fig. 5.14. First, a prism is set over Point 4. The total station is set over Point 1. A second prism is set over Point 2. The order of the setups does not matter. Before the angle and distance can be measured, it is necessary to record two point numbers in the data collector. In this case, we tell the data collector that the backsight is point number 4 and the occupied point is 1. This information is stored in the raw data file.

In order to turn the angle, the instrument person first sights point 4. As we have seen, total stations have knobs or tangent screws which can be used to precisely point the telescope in both the horizontal and vertical axes. When the instrument person is comfortable that Point 4 is sighted as precisely as possible, the total station horizontal circle is set to zero. This locks zero on the backsight so that when the total station is subsequently turned to sight another point, the angle can be read off the total station. There is usually a button on the total station or data collector for that purpose.

The total station is then turned and Point 2 is sighted. The total station or data collector is then told to measure the distance to point 2. Once the measurement is

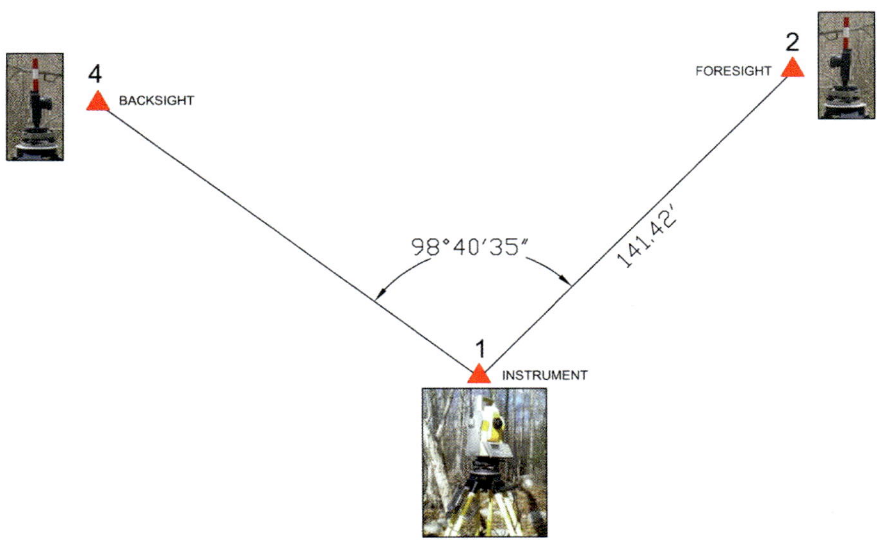

Fig. 5.14 Turning an angle with a total station

taken the information is stored in the memory of the data collector. The data collector will prompt the user to enter the name and description of the foresight—in this case the name will be Point 2. We could also add a description of point 2, such as *Old Stone Bound Found* or whatever description best fits the point. The data collector will then store the horizontal angle, slope distance, vertical angle, point number and description. In this case the horizontal angle would be AR98° 40′ 35″. Notice that the horizontal angle was turned to the right, which is why the angle is preceded by AR. The slope distance would be 141.42 feet. The point number would be 2. If the angle were being measured to a traverse point or a point where the best possible accuracy was desired, the angle would be turned multiple times and the results averaged.

If this were a traverse, the total station would be moved to Point 2 and the backsight would be set on Point 1. A new point would be established ahead of Point 2 and the process would continue. Because tribrachs allow the equipment to be interchanged on the same tripod this procedure is sometimes referred to as **Leap Frogging**.

5.16 Example of Using a Total Station to Set an Angle and Distance

Using a total station to set an angle and distance is a common procedure when setting boundary corners. An example is shown in Fig. 5.15. A traverse line, established during the initial field work, exists between points 1 and 2. All calculations have been performed and it is now necessary to go back into the field to set

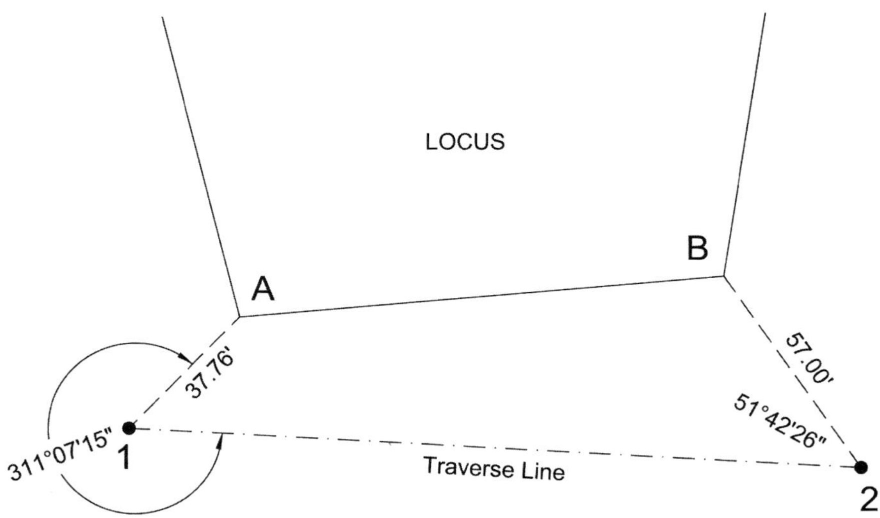

Fig. 5.15 Setting lot corners

lot corners A and B. The angles and distances from the traverse line to the lot corners were calculated in the office and noted on the sketch shown in Fig. 5.15. The sketch is taken into the field so that the crew will have the information necessary to set the lot corners. When using a data collector, the coordinates for points 1, 2 and A, B will be stored in the data collector, so they can simply be called up in the field and the data collector will calculate the required angle and distance needed to set the lot corners.

The procedure used for setting angles and distances is similar to the previous example. The difference is that the angle and distance to the foresight are already known. Let us first set lot corner A. The total station is set up on point 1. The backsight is set on point 2. The instrument person then sights point 2 and sets zero on the total station. (Setting zero on the backsight is the same as setting 360° on the backsight.) Next, the total station is turned until the angle display reads 311° 07′ 15″. Note that the total station needn't physically be turned all the way around to the right. It can simply be turned to the left and the angle will still be measured clockwise to the right. In effect, as the telescope is turned to the left the angle will begin to decrease from 360°. The total station telescope is now pointing toward lot corner A. The instrument person then directs the rod person until she is holding the prism pole on line. Initially, the rod person could guess at the 37.76 foot distance from the total station or she could pace the distance off from the total station. She could also use a tape to measure the distance. Modern EDMs are able to measure a distance in 2 or 3 seconds so it is common to just guess at the distance or pace it off. An EDM distance can then be taken and the rod person can move in or out until the exact point is located.

Rod persons carry, or should carry a small retractable tape similar to the ones available at any hardware store and commonly used in the carpentry trade. The difference between the carpentry version and the survey version is that the latter is graduated in tenths and hundredths of a foot instead of inches. For setting out points, the tape is very useful and can greatly reduce the number of EDM shots required. Once the rod person is within a foot or two of the point to be set, the tape can be used to measure the exact distance to the point. Using our example, let's say the rod person is adept at pacing so the first EDM distance is 36.56 feet. The rod person can use the tape to measure the remaining 1.20 feet exactly. The next EDM shot should confirm that the prism pole is in the right location. Alternatively, if a tape was not used and the rod person guessed the distance each time, it could take a dozen shots before the rod person arrived at the correct point. This is a waste of time and a waste of the client's money. Although a skilled rod person might be able to arrive at the correct location in a half dozen shots, using a tape will almost certainly save time. Keep in mind that the lot corner will probably need to be within 0.01′ of its calculated position. It is extremely difficult to consistently guess with this level of precision.

Once point A has been set, the total station would be moved to point 2 and the backsight to point 1 where lot corner B could be set. If tripods and tribrachs have been used for the setups at points 1 and 2 it would only be necessary to swap the prism and the total station so no actual setups would be required. This time the angle to be turned would be 51° 42′ 26″ to the right. The same procedure described above would be used to measure out 57.00 feet from the total station and set the corner. Finally, the distance between points A and B should be checked. This could be done by using a tape if the distance is relatively short. Alternately, the total station could be set on A or B and the EDM used to check the distance. If neither of these alternatives is possible, the angle and distance from 1 to B and 2 to A could be calculated and the total station used to measure the angles and distances to the points.

Setting lot corners is one area where a servo driven tracking total station really pays for itself, particularly when there are many lot corners to set. Once the total station and backsight are set up, the data collector is used to specify which point needs to be set. The data collector will then cause the total station to automatically turn in the direction of the point. Once the rod person is near the point, the total station can be set to track the rod person and continually update the distance and angle measurements. Listening to instructions from the instrument person, the rod person can essentially walk to the lot corner. This is even easier with a robotic total station or RTK GPS because the rod person would then have the data collector—usually attached to the prism pole. In this case the rod person can read off the distance correction herself. For example, the data collector for a robotic total station might display "Move 0.15′ Left and Go Out 0.67′". Graphic data collectors will also have an image of the point to be set and the location of the prism pole. The person setting the point can then graphically see which way to move and how far to move.

5.17 Some Commonly Used Surveying Techniques

Double Centering. All surveying instruments have small errors which can affect the accuracy of a survey. Instruments are calibrated to eliminate errors but sometimes small errors continue to exist. Let us consider an example where we need to extend a straight line. Imagine that you are laying out a new airport runway and you have been asked to place marks on the runway so that a centerline can be painted. You only have two points. One point is in the middle of the pavement at the approach end of the runway. The second point is in the middle of the pavement halfway down the runway. This is shown in Fig. 5.16. You need to establish a point at the departure end of the runway.

You set up the instrument at the halfway point. You sight a point at the approach end then, without touching the horizontal circle clamp or tangent screw, you invert the telescope and sight the departure end of the runway. You have

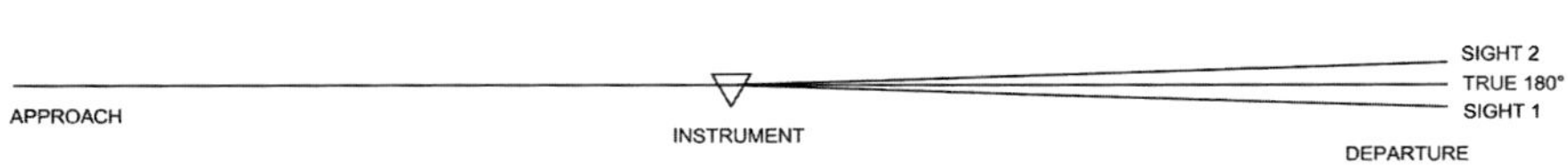

Fig. 5.16 Double center

an associate make a mark on the runway at the exact point which you have just sighted. This point is shown as Sight 1 in Fig. 5.16. The point which was just set on the runway should be exactly 180° from the point at the takeoff end. In other words, there should be a straight line between the first point sighted, the instrument and the point just set.

In order to check the accuracy of the straight line you would loosen the horizontal circle clamp, turn the instrument 180° and sight the same point at the approach end of the runway with the telescope inverted. The telescope would then be flipped over to its erect position and your associate would mark another point on the runway. This point is shown as Sight 2 in Fig. 5.16. If the two points are at the same location, the instrument is perfectly calibrated and you have a straight line. If they are not, the instrument contains an error. In practice, most instruments will contain a small error. In our example, Sight 1 and Sight 2 are not the same point so our instrument is not properly calibrated. We desire a point which would give us a straight line between the approach end and departure end of the runway. This point will be halfway between Sight 1 and Sight 2. The procedure we just described is called **double centering**. An alternative to double centering is to turn 180°. When turning an angle, the accuracy is subject to the precision with which an angle can be turned. This is not a factor when double centering so it is the preferred method for projecting a line.

Turning Multiple Angles. In order to increase the precision of traverse measurements it is common to turn multiple angles and average them. Sometimes either the backsight or foresight is not sighted perfectly. Performing multiple measurements will tend to reduce these errors and increase the precision of the angle being measured. In the example shown in Fig. 5.14, we measured the angle 4-1-2 with the result of 98° 40′ 35″. We could turn this angle a second time by locking the resulting angle into the total station display, sighting the backsight again and measuring to the foresight a second time. If all of our sightings were perfect, the resulting angle will be double the value of the first angle. Because sightings are rarely perfect, there is usually a small difference between the two measurements. In our example, we might read the second angle as 197° 21′ 20″. If we divide this result by 2 we find that the average angle is 98° 40′ 40″. If we require even more precision we could continue the process and turn the angle more times. We saw in our runway example that the instrument may not be perfectly adjusted. To account for this it is best to turn our multiple angles with the scope both erect and inverted, as this corrects for misadjustment.

Wiggling In. Occasionally, a surveyor needs to establish a point on a straight line between two points. In most cases, this is not too difficult. The instrument can be set on the first point and the second point sighted. The intermediate point can then be placed on line. Suppose, however, that the two points are separated by a hill so that the two points are not intervisible (one of the points cannot be seen from the other point)? This situation is shown in Fig. 5.17, where a hill prevents sighting between the two markers, 1 and 2, at the bottom of the hill. Surveyors get around this problem by setting the instrument at the top of the hill and sighting each monument. The problem is that there is no way to know if the instrument is

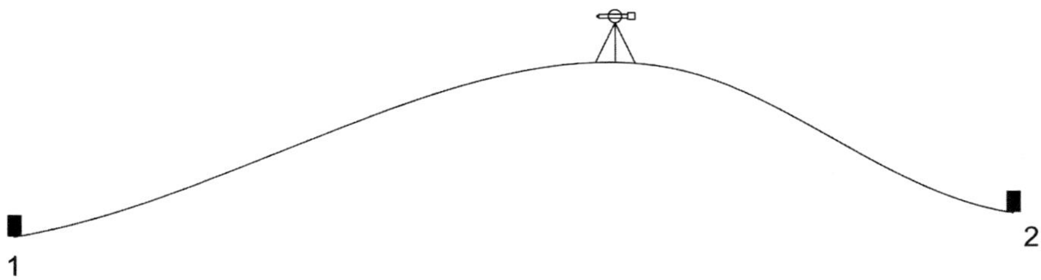

Fig. 5.17 Wiggling In

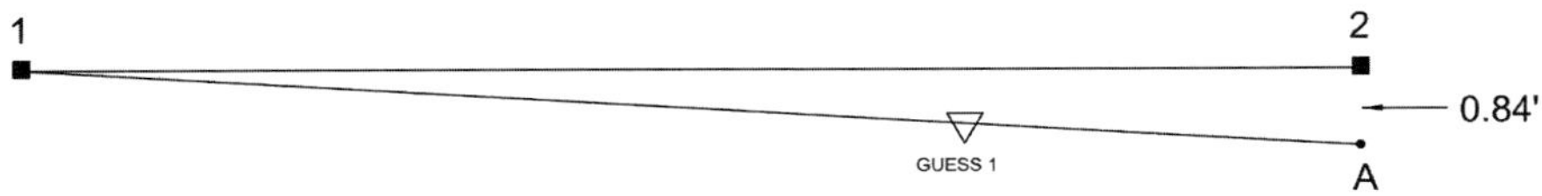

Fig. 5.18 Wiggling In—pick a point

on a line between the two points. As there is no frame of reference, it is very difficult for most people standing between two points to determine whether or not they are close to being on line.

In spite of these limits, we will do our best to guess at the location of the intermediate point. An example of how this might look is shown in the overhead view in Fig. 5.18. The line between point 1 and point 2 is the line that we need to establish. It is a straight line between the two monuments.

The point Guess 1 has been selected as a trial location for the instrument. The instrument is set up on point Guess 1 and point 1 is sighted. We always want to sight the long line rather than the short line because projecting a line from a long line to a short line will not magnify an error as much. The instrument is then turned to an angle of 180°. A point is then set near point 2. In our example we will call this point "A".

Next, we measure the distance between A and point 2. We can see from Fig. 5.18 that our point is 0.84 feet away from point 2. If our instrument were halfway between points 1 and 2 it would be a simple matter to move the instrument 0.42 feet—half the error. The next time we turned 180° we should be right on point 2. In our example, we are about 2/3 (0.67) of the way between 1 and 2 so we would want to move the instrument about 0.56 feet over (0.67 * 0.84 = 0.56). After we moved the instrument over, we would sight point 1 again and turn 180°. If we were still a little bit off point 2, we would measure the difference again and readjust the instrument location. We would keep "wiggling in" as many times as necessary until the instrument was exactly on line. In this example, we turned 180°, however we could have also double centered point A and used the averaged result of the erect and inverted shots to establish point A.

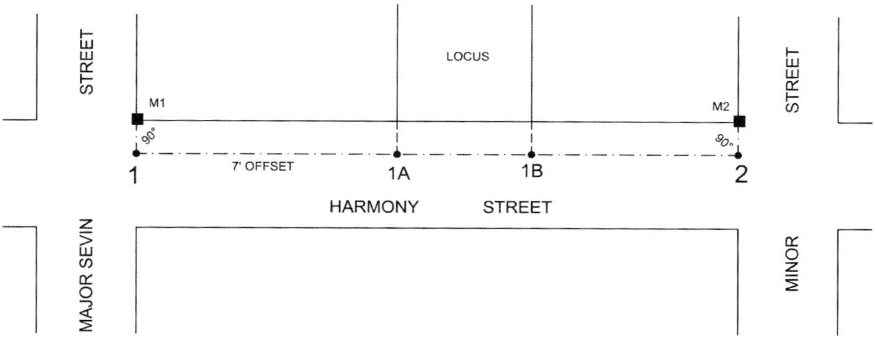

Fig. 5.19 Wiggling In—offset line

A more complicated example of wiggling in occurs when we need to establish an offset line between two monuments that are not visible from each other. Knowing how to do this is very useful when working in urban or suburban locations where most streets have monuments at the intersections. One might think it would be easier just to set up on one of the monuments and sight the other one, but there are usually fences or hedges in the way which make sighting impossible. An example of a 7 foot offset line is shown in Fig. 5.19. It is the dashed line running between points 1 and 2. Notice that there is a monument at the intersection of Harmony St. and Major Sevin St., shown as a small black square M1. Another monument exists at the intersection of Harmony St. and Minor St. labeled M2. These are the two monuments that we wish to offset.

If a surveyor was required to establish the lot corners of the parcel labeled **Locus**, the 7′ offset would allow working on the sidewalk, outside of obstacles which might exist on the street sideline. Once the offset line was established, the surveyor could measure along the offset line to points opposite each of the lot corner and set points in the sidewalk—for example from point 1 to point 1A and 1B. Points 1A and 1B are called **Points on Line** because they both lie on a straight line—the line from 1 to 2. Both points also lie on projections of the property lines, so once the instrument was set up on 1A or 1B it would be possible to sight down the property line and set both the front and rear lot corners (providing there were no obstructions on the property line). Short dashed lines are shown in Fig. 5.19 running from 1A and 1B to the property corners of the Locus.

The offset line from 1 to 2 is established by wiggling in. Wiggling in to establish an offset line is a bit more complicated than wiggling into a straight line between two points, as in our previous example, but it is easily accomplished nonetheless. A surveyor would proceed as follows. A tape would be used to measure out 7 feet from each of the monuments. Points would be set in the sidewalk at the 7 foot offsets. In our example, we need to be exactly 7 feet off of the street line, so the 7 feet must be measured perpendicular to the street line. The difficulty lies in establishing points 1 and 2 at 90° to the street line. This is usually done by eye or by using a small handheld right angle prism.

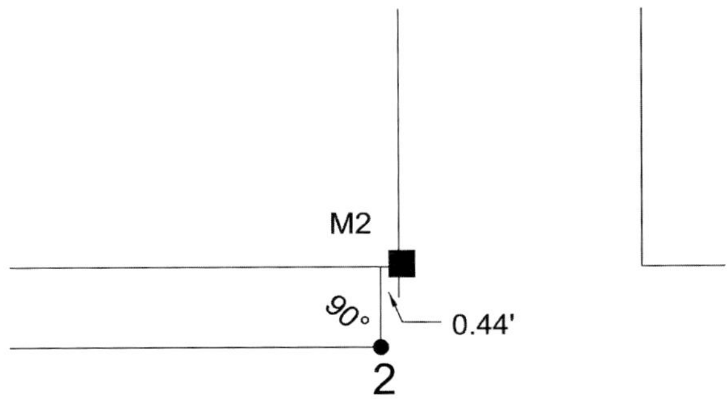

Fig. 5.20 Wiggling In—close up of monument near point 2

Once points 1 and 2 have been marked on the sidewalk, the instrument would be set up over point 2 and a target set over point 1. The instrument would then set zero on the target and turn 90° right. If the location of point 2 was guessed correctly, the M2 would be on line. It usually isn't—at least not exactly. If point 2 was not guessed correctly, line would be on one side or the other of the monument. Figure 5.20 is an enlarged view of the area near point 2. In our example, we can see from Fig. 5.20 that we missed M2 by 0.44 feet. It will therefore be necessary to move point 2 0.44 feet to the right. The reader must keep in mind that when an offset is first measured with a tape, if the point to be established is not exactly 90° to the street line the offset distance will be shorter than the desired distance. In our example we measured 7 feet from M2 in order to establish point 2. After turning the angle with the instrument we discovered that we were 0.44 feet off line. This means that the 7 feet was measured at an angle that was not 90° to the street line so point 2 will be a little closer to M2 than 7 feet. That is why it is a good idea to recheck the offset dimension whenever an offset point needs to be moved. Once we have moved point 2 and have confirmed that the angle from 1-2-M2 is 90°, we would then set the instrument on point 1 and repeat the process of wiggling in to M1. When establishing an offset line, we must wiggle in at each offset point, so it is twice as much work as wiggling into a straight line.

Chapter 6
Road Geometry

Most parcels of land have frontage on public or private roads or streets so it is beneficial to spend a little time reviewing typical road geometries. Calculating road geometry requires an understanding of geometry and trigonometry so a comprehensive treatment of the subject is beyond the scope of this portion of the book. Nevertheless, the explanations found in this chapter will provide the reader with a good foundation for understanding the geometry of roads and road curves. For those wishing a more comprehensive discussion, the subject is treated in greater depth in the appendix.

When moving automobiles change direction they do not do so instantaneously. Direction is changed gradually by going around a curve. On a highway, where vehicles are traveling at high speed, the curves are gradual so they can be negotiated without changing speed. In the city where travel is slow, (or should be) and streets intersect at right angles, curves are relatively abrupt or even non-existent. The feature that makes a curve gradual or abrupt is known as the **Curve Radius**. The longer the curve radius, the more gradual the curve is and it can be negotiated at higher speed. The shorter the curve radius the more abrupt the curve is and it must be negotiated at lower speed. The names **Road** and **Street** mean the same thing in our context so the names will be used here interchangeably.

Roads are often described using the term **Road Layout**. A road layout is a plan that describes the features of a road, such as road dimensions, intersections, curve geometry, grades and other attributes. Layouts are often held in public recording offices such as highway engineering departments and recording offices which hold plans and deeds. In some cases, when roads are newly created, widened or relocated, engineers working for the municipality or state will design the road, create plans, lay it out on the ground and set monuments which fix the limits of the road boundaries. Often, new road layouts must be approved by the community or its representatives before it achieves legal status. This is always the case when a new or revised layout requires the taking of private property.

P. Gay, *Practical Boundary Surveying*, DOI 10.1007/978-3-319-07158-9_6

6.1 Curves in Roads and Highways

All curves used in road layouts share a similar geometry. This geometry is shown in Fig. 6.1. The geometry in our image is based on a road that is 40 feet wide. All of the road sidelines on the north side of the road are parallel to those on the south side of the road. Road sidelines are usually referred to as **Street Lines**. The left-most vertical red dashed line labeled 100.00′ (beginning at the label **PC**) shows the **Radius** of the inside curve. The radius is usually labeled **R**. The radius of the outside (north side) curve is not shown but it would be 140.00 feet (100′ + 40′). Notice that the radius line is at right angles (90°) to the sideline of the road. Surveyors would say that the curve is **Tangent** to the street line.

Notice that the red dashed line at the end of the curve also makes a 90° angle with the sideline of the road. The curve is tangent at this point also. In fact, almost all road curves are tangent to the sidelines at the beginning and end of a curve. Curves can therefore be assumed to be tangent unless they are specifically labeled as non-tangent. The curve on the north side of the road is also tangent at both ends of the curve. It has to be because both the north and south street lines are parallel to each other.

The beginning of the curve is labeled **PC**, which means **Point of Curvature**—the place where a curve begins. The end of the curve is labeled **PT** which means **Point of Tangency**—the place where a curve ends. The fact that the names are different can be a bit confusing because the curve is actually tangent at both places. So why aren't they both labeled PT? The reason is that when making road layouts it is

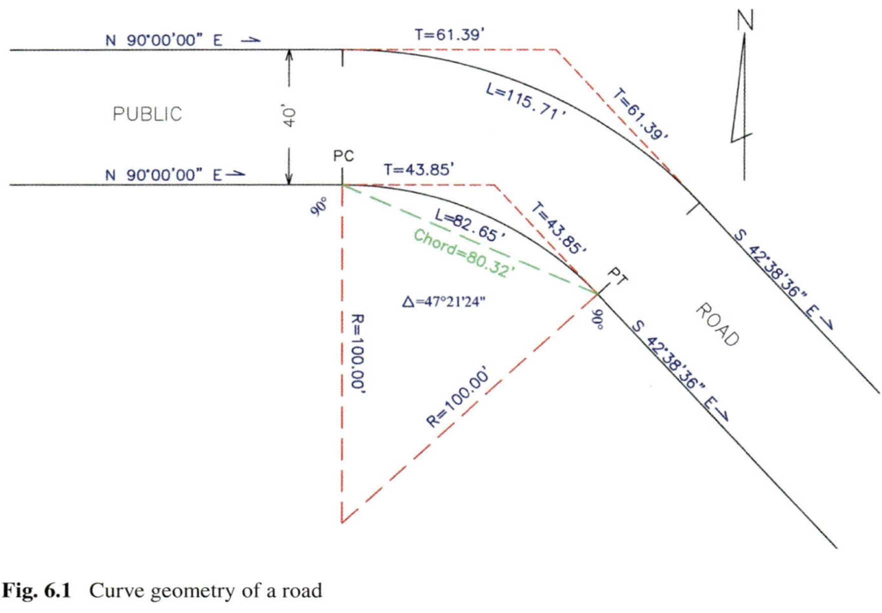

Fig. 6.1 Curve geometry of a road

customary to show bearings on the street lines. As we have already learned, bearings indicate a specific direction. In our example, the street begins at the left side of the page with a bearing of N90°E (due east) and ends with an angle of S42°E. In other words, the layout in our example is described from left to right. The label PC shows where the curve starts and the label PT shows where the curve ends.

It is customary to set permanent monuments at the PC and PT of every curve in public street layouts. Normally these will be either stone or concrete bounds.

The length of the curve on the south side of the street is 82.65 feet. In street layouts, the curve length is sometimes labeled as **L** or **LC**. The **Delta**, labeled with the Greek symbol Δ, is the angle between the two 100 foot radius lines. It is also the angle of the *bend* or *deflection* in the street. For example, if you were driving down this street starting from the left side of the page, you would be initially travelling in a straight line. When you got to the curve (PC), you would gradually turn your vehicle to the right. When you got to the end of the curve (PT) you would be driving in a straight line again. The angle through which your vehicle turned (about 47°) would be the same as the curve delta. If a street made a right angle turn, the delta would be 90°. If the angle in a street were gradually diminished so that the curve went away and the street became straight, the delta angle would be reduced until it was finally 0°.

Another piece of information commonly shown on curves is the **Tangent** which is labeled **T**. The tangent on the south side of our example street is 43.85 feet. This is simply the distance that the two straight sidelines would have to be extended in order to intersect. Tangent distances on a curve are always equal (unless the curve is non-tangent to the street line or there is a compound curve). Tangents are useful for surveyors when laying out streets because an instrument can be set up at the point where the tangents intersect and sight down the street line in both directions. However, this has become less important as computers and data collectors have made calculations so easy that where you set up the instrument doesn't matter very much anymore.

The last piece of information shown is the **Chord** length, labeled **C** or **chord**, of 80.32 feet. The chord is the straight line distance between the PC and PT. Knowing the chord is handy because if there are monuments at the PC and PT, it is a simple matter to measure the distance between them with a tape to see if they have been disturbed. The chord is also handy when one is looking for the monuments in order to establish the street lines and no instrument is available. If one of the monuments is visible but the second one is buried, using the chord distance to measure from the visible monument will help a surveyor know where to look for the second one.

6.2 Intersecting Streets

Let us now consider street geometry which would be typical of streets intersecting in an urban location. The intersection under consideration is shown in Fig. 6.2. The geometry is fundamentally no different than in the previous example. In this

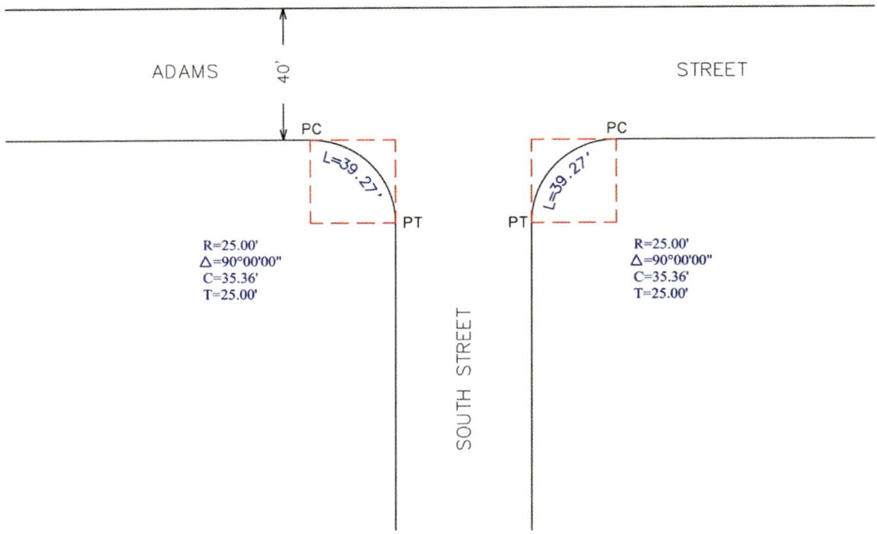

Fig. 6.2 Curves at street intersection

case the streets intersect at 90°. The curve radius is 25 feet. Notice that the tangent is exactly the same as the radius, 25 feet. Because the delta is 90° the tangent and radius have to be the same length. The curve length is 39.27 feet. For readers who recall how to calculate the circumference of a circle, the cord length shown is exactly ¼ the diameter of a circle having a radius of 25 feet.

In this case, there are no bearings on the street lines so the labels PC and PT are arbitrary. It really doesn't matter as long as PC is used for one end of the curve and PT for the other. All of the information presented above relating to curve geometry can be calculated as long as two of the values are known. One of the values must be the radius or delta angle.

Chapter 7
The Public Land Survey System

The **Public Land Survey System** (PLSS), also known as the **Rectangular Survey System**, is a method of dividing land into rectangular parcels. The system began in 1785 in Ohio, and was further established by the General Land Office (GLO), an independent agency of the U.S. Government, during the 19th century. The GLO was eventually merged with the Bureau of Land Management (BLM) when it was created in 1946.

The system applies to 30 states, as shown by the colored states in Fig. 7.1. States which did not adopt the system are, Georgia, Connecticut, Delaware, Kentucky, Maine, Maryland, Massachusetts, New Hampshire, New Jersey, New York, North Carolina, Pennsylvania, Rhode Island, South Carolina, Tennessee, Vermont, Virginia, and West Virginia and Texas (Texas has a variation of the PLSS system).

7.1 The Manual of Surveying Instructions

Surveys made under the PLSS are governed by the manual prepared by the Bureau of Land Management. The book is available from the BLM and the current version is entitled *The 2009 Manual of Surveying Instructions* (referred to herein as Manual). The first manual was published in 1855 and there have been 9 editions. The Manual contains more than 300 pages so, with our limited space, all we can hope to do here is to provide the reader with a broad overview of the PLSS.

7.2 Principal Meridians and Base Lines

The system is based on established **Principal Meridians** and **Base Lines** as shown in the map, published by the U.S. Government and shown in Fig. 7.1. These principal meridians and base lines start from **Initial Points**. Shown at the intersection of black lines on the map, the initial points are fixed on the ground at

P. Gay, *Practical Boundary Surveying*, DOI 10.1007/978-3-319-07158-9_7

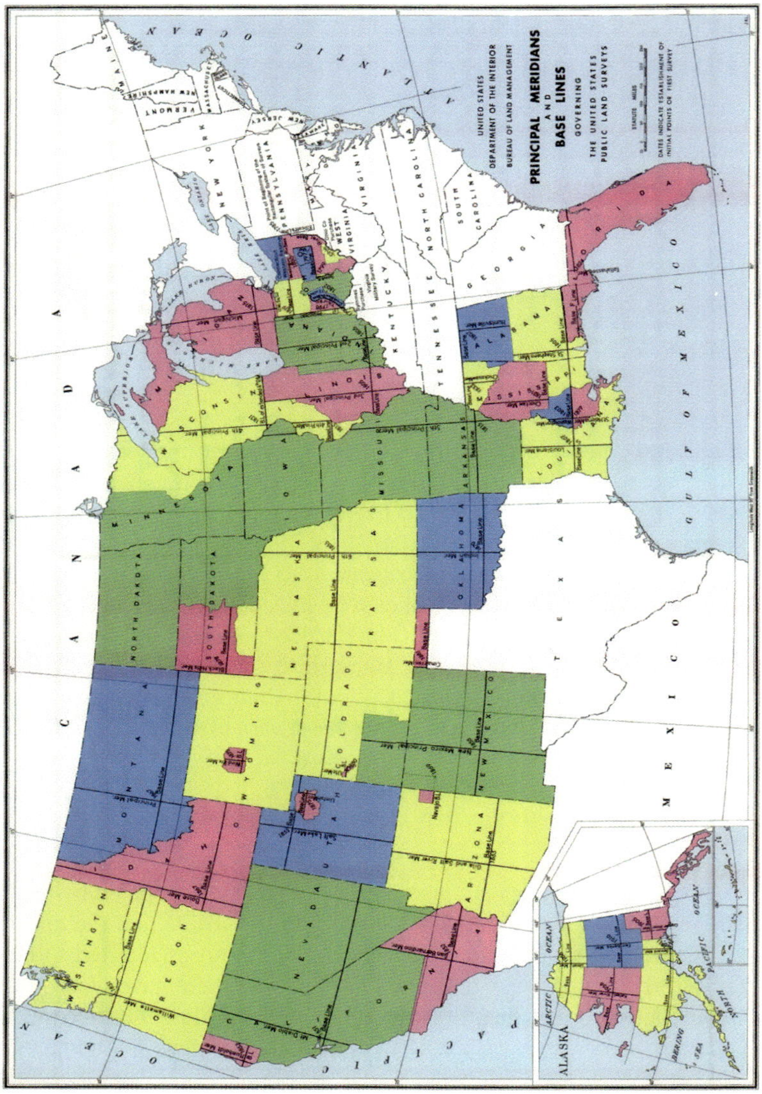

Fig. 7.1 Map of the principal meridians, base lines and initial points published by the U.S. Government

a number of locations in the U.S., such as California, Oklahoma, Wyoming and other locations. The principal meridians and base lines originate at these points. The Manual contains a listing of the meridians and base lines for the various locations in the U.S.

Meridians are north–south lines which converge at the earth's poles. The earth is approximately spherical. All meridians have their greatest spacing at the

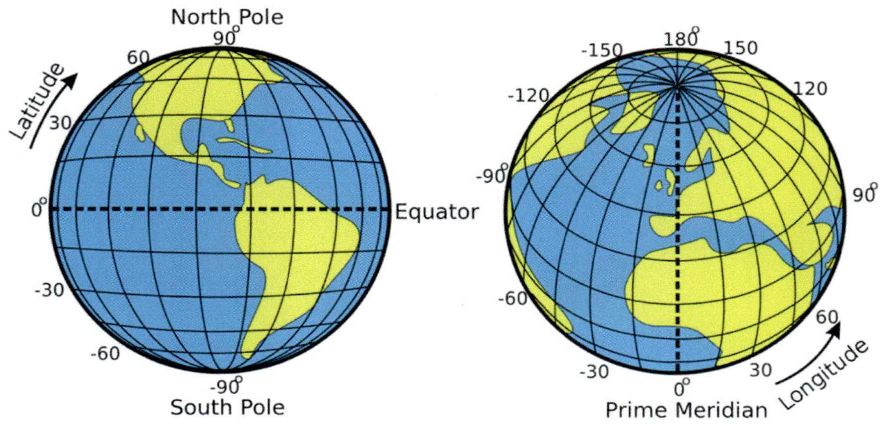

Fig. 7.2 Parallels and meridians

equator. As the meridians progress north or south from the equator, they converge and come to a point at the poles. This is called **Convergence**. Figure 7.2 shows parallels and meridians. As shown in Fig. 7.1, principal meridians were established which pass through initial points. Principal meridians extend both north and south from these points. All principal meridians run true north and south.

Base Lines are parallels of latitude. Parallels are so named because they are parallel to the equator. Unlike meridians, Parallels do not converge. As shown in Fig. 7.1, base lines are extended east and west from the initial points. The intersection of a base line and a meridian forms a 90° angle. Base lines run true east and true west.

Notice in Fig. 7.3 near R4E that south of the base line, where the first guide meridian east intersects with the base line, there are two intersections. The intersection to the south of the base line is called a **Closing Corner**. The closing corner is the intersection of the meridional line with the parallel. The intersection to the north of the base line is called a **Standard Corner**. The offset in the two corners is caused by convergence of the guide meridians and occurs on each standard parallel.

7.3 Quadrangles

Once the principal meridians and base lines were established, a series of **Standard Parallels** were established 24 miles north and south of the base lines (in some cases 30 miles was used). The labeling was sequential, such as first standard parallel north, second standard parallel north and first standard parallel south, second standard parallel south. An example of these parallels is illustrated in Fig. 7.3.

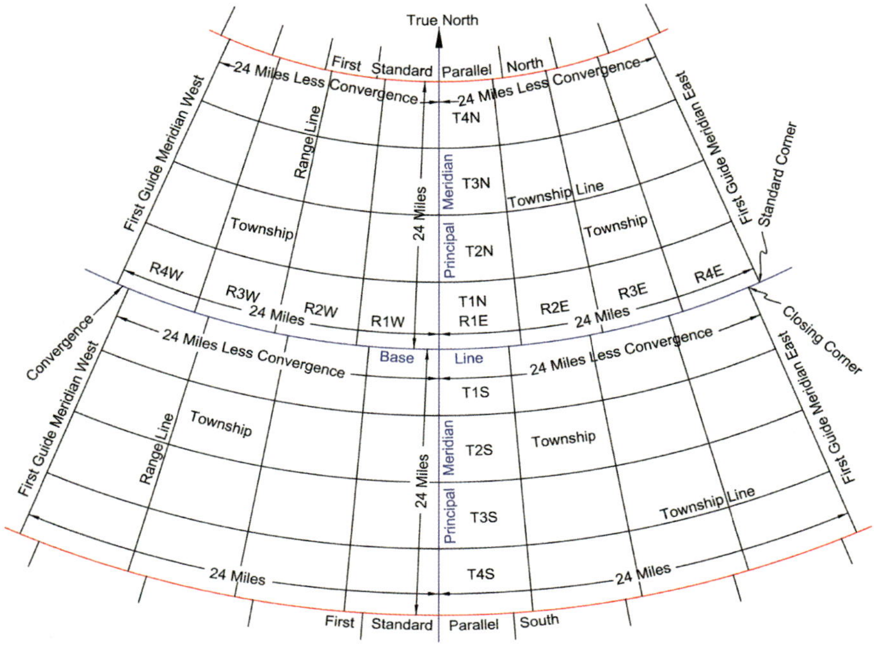

Fig. 7.3 Quadrangles. four quadrangles are illustrated

Meridians were then established 24 miles east and west of the principal meridian. These were called **First Guide Meridian East** and **First Guide Meridian West**. The standard parallels and guide meridians created quadrangles 24 miles on each side.

7.4 Townships

Each of the quadrangles was then divided into 16 **Townships**. Townships are approximately 6 miles square. The townships are bounded by **Range Lines** which run north and south and **Township Lines** which run east and west. The meridians and parallels conform to the earth's curvature. So do the range lines and township lines. These features can be seen in Fig. 7.3.

When townships were created, the dimension of 24 miles along the base line was held. Because of convergence, the range lines will become closer together the more northerly they become. Consider the range lines beginning at the base line and running north to the first standard parallel north. When the first standard parallel north is reached, the convergence will be maximum for that particular quadrangle. This means that the east-west dimensions of the townships will be reduced as they approach this parallel. The quadrangle distance along the first standard parallel north will be less than 24 miles. That is why it is labeled "24 miles less convergence".

The same situation exists for the first standard parallel south. The township boundaries along this parallel will be 24 miles. However, when the base line is reached, convergence will cause the dimension of the quadrangle to be less than 24 miles. Notice in Fig. 7.3 that the range lines of the two abutting quadrangles do not intersect at the base line. The township widths immediately south of the baseline are less than those which are north of the baseline. This convergence becomes important when we work with sections.

7.5 Sections

After the townships were created, each was divided into 36 **Sections**. A section is one mile square—at least some of them are. Each section contains 640 acres. Figure 7.4 shows a township divided into sections.

The convergence of the range lines, which fix the township boundaries, makes it impossible to create 36 sections which are exactly square. This is a common problem in surveying. It occurs when a two dimensional plane must be fitted to a spherical earth. There are a number of ways this problem might have been solved. One way would have been to alter the shape of each township so that 36 of them would fit exactly in a township. However, this would make surveying difficult as the sections would no longer be squares. Another way to deal with the

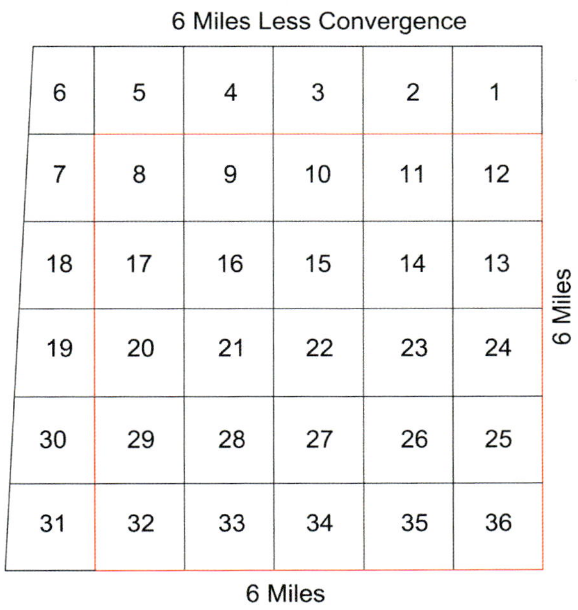

Fig. 7.4 A township divided into 36 sections. Sections outlined in *red* are one mile square. Sections on the north and west boundaries are subject to dimensional changes resulting from convergence

problem would be to make most of the sections square and place the ones with the odd dimensions on the outside boundary of the township. This was the method selected.

Sections were divided from townships by starting at the northeast corner of the township with section 1, then going west to section 6. Section 7 starts south of section 6 and the numbers then increment easterly. This process is continued with numbers in each row increasing alternately to the east and west until section 36 is reached at the southeast corner of the township.

Starting at the southeast corner of the township and running north and west, all of the sections are one mile square, except for the sections along the north and west borders of the township. Sections 8–12, 13–17, 20–24, 25–29 and 32-36 are all dimensioned as one mile squares. These are outlined in red in Fig. 7.4. The dimensional changes resulting from convergence are placed in the sections bounding the northerly and westerly boundaries of the township. The sections subject to dimensional changes are 1–7, 18, 19, 30 and 31.

7.6 Nomenclature for Locating Townships and Sections

Division of a section is shown in Fig. 7.5. This may seem a bit confusing at first but it is actually quite logical. For example, the NW ¼ is the northwest quarter of the section. The N ½ NE ¼ is the north half of the northeast quarter. One of the smallest divisions shown, comprising 10 acres, is labeled NW ¼, SW ¼ SE ¼ which translates to the northwest quarter of the southwest quarter of the SE quarter. So, you would first find the SE quarter section, find the SW quarter of the quarter section then find the NW quarter of it.

The location of townships and sections is made using specific descriptors. An example will best illustrate the method. Consider the following: T2NR5 W S4 NMPM. This would read: Township 2 North, Range 5 West, Section 4, New Mexico Principal Meridian. Each principal meridian has a standard abbreviation. Equal divisions of sections such as quarter and half sections are called **Aliquot Parts**.

7.7 Measurements Under the PLSS

Under the PLSS, the basis of all distance measurements is the Gunter's chain. The chain is 66 feet in length. It has 100 links so each link is 0.66 feet. One mile is 80 chains or 5,280 feet. All measurements are horizontal. Of course, no one uses a physical chain any more. However, it is important to understand that the basis of modern measurements under the PLSS is the chain.

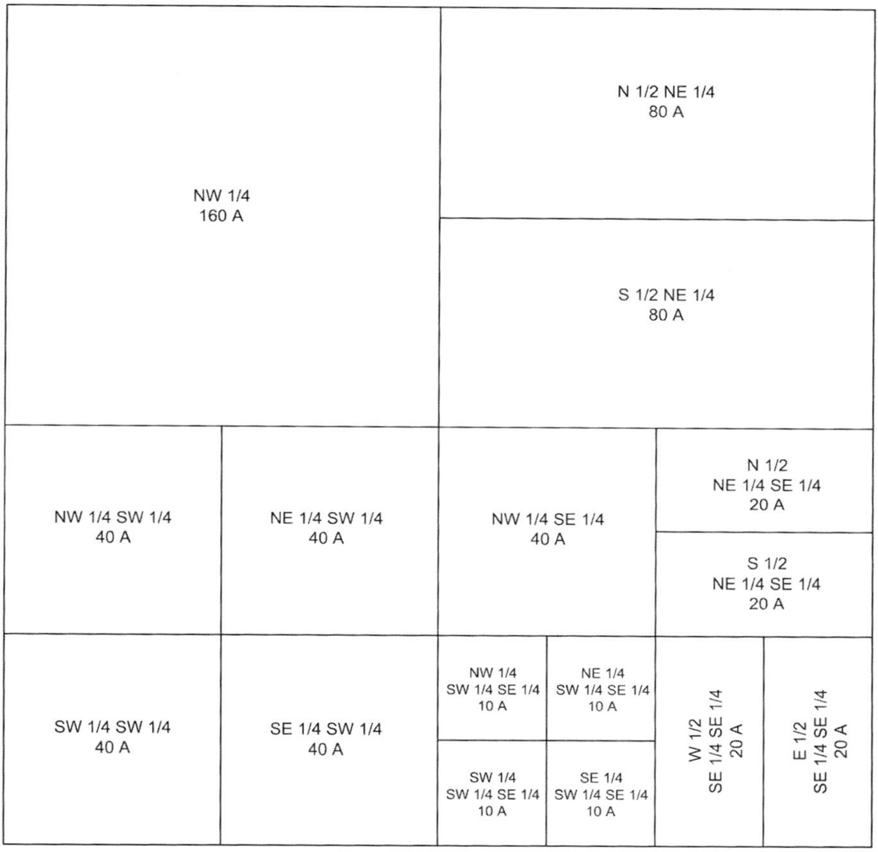

Fig. 7.5 Nomenclature for division of a section

7.8 Monumentation Under the PLSS

The corners established during the original surveys are forever fixed by law. If there were errors made during the location of the original monuments, and in some cases there were, the location of the monument prevails. The original position of the monument carries far greater weight than courses and distances.

The terms **Corner** and **Monument** do not mean the same thing. A corner is the point on the ground determined by the original survey. A monument is the object or physical structure which marks the corner point. Even if the physical evidence of a corner may have disappeared, the corner will not have been lost if its location can be determined by some other physical evidence, by field notes or by testimony. Under federal law, it is a criminal offense to destroy, deface or change the location of a Federal monument.

The BLM has regulations for the types of monuments used in public surveys. Although a surveyor may be authorized to utilize other types of monuments, the **Regulation Post** is an alloyed iron pipe, zinc coated, 2 ½ inch. in diameter and cut to a length of 30 inch. The bottom of the pipe is split and spread. A brass cap is affixed to the top of the pipe. Brass tablets are specified for installing in rock where it is not possible to use the pipe. Monuments are marked for identification of the corner. The Manual goes into detail on the form and abbreviations to be used for marking monuments.

Many old monuments obviously do not conform to the present BLM requirements. Many of the older monuments were piles of rocks or piles of rocks with wooden posts. Even trees could be marked and used if they happened to coincide with a corner location.

Witness Corners are monuments which are usually placed on a survey line near a corner. Witness corners are used when it is not practical to place a monument on the corner itself. For example, the corner may fall in the middle of a stream. It is preferred that the witness corner be located within 10 chains of the actual corner.

7.9 Meandering

All navigable bodies of water and important lakes and rivers are distinct from public lands. The mean high water line is the boundary between these bodies of water and the public lands. The high water mark is defined as the line which the water impresses on the soil by covering it for sufficient periods to deprive it of vegetation. A **Meander line** is a traverse consisting of straight line segments which is run along the high water mark (Fig. 7.6). Section lines are run to the meander line and the intersection is called a **Meander corner**. The meander line therefore begins at a section corner and ends at the next section corner (or township corner). All lakes having an area of 50 acres or more are meandered. Navigable rivers, bayous and rivers which are not navigable but which are at least 3 chains across are meandered on both banks. Shallow and intermittent streams are not meandered even if they are more than 3 chains across. It is important to understand that the meander lines are used for the purpose of calculating area. They are not an actual boundary. The stream or other body of water is the actual boundary.

7.10 Restoration of Lost Corners

Restoration of lost corners under the PLSS is closely regulated by the Manual of Instructions for the Survey of the Public Lands of the United States. This Manual is in excess of 300 pages so it is only possible to cover a fraction of the material here. We will therefore only be able to consider a few of the examples of the restoration of lost corners.

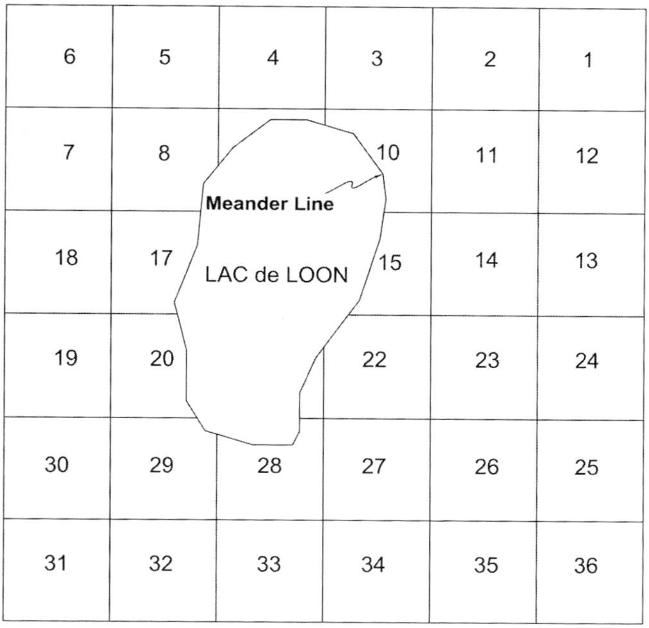

Fig. 7.6 Meandering around a lake meander

Restoration of lost corners under the PLSS is closely regulated by the Manual of Instructions for the Survey of the Public Lands of the United States.

The boundaries of public lands are fixed and cannot be changed. The locations of the corners marking the division of land are identified by monuments, descriptions in the field notes of the original survey, supplemental evidence and by testimony. Field notes contain information from the original survey of the land. The field notes are as important in the grant as the original plat which shows and describes the property. Even if an original monument has disappeared it will not be regarded as lost if there is testimony from a witness who has reliable knowledge of the original location. An **Obliterated Monument** is a monument where there is no remaining physical evidence of the monument but there is evidence of its location from the testimony of witnesses, surveyors or other credible witnesses.

A **Lost Corner** exists where the location of the corner cannot be determined from physical evidence or reliable testimony, and the location of the corner can be restored only by reference to other corners. The rules for restoring a lost corner must not be applied unless there is no other way to replace the corner. When it becomes necessary to replace a lost corner, the replacement must be done by **Proportionate**

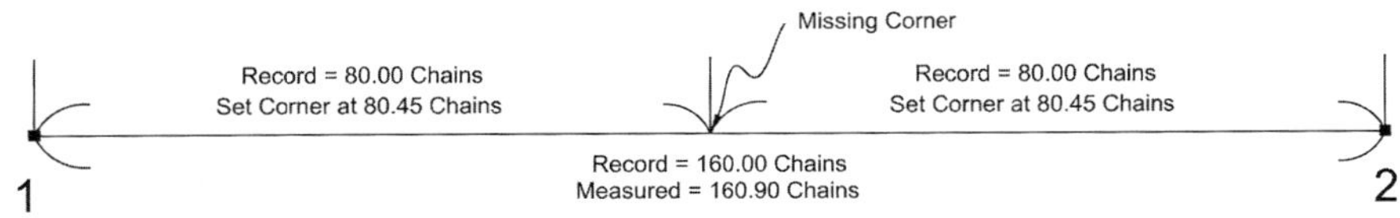

Fig. 7.7 Single proportionate measurement

Measurement. Proportionate measurement gives equal relative weight to all parts of the line. The ends of the lines are fixed and the line is divided such that each length is adjusted in proportion to the excess or deficiency in the entire line.

7.11 Single Proportionate Measurement

Single proportionate measurement is used when a corner is missing on a line. The position of the missing corner is based on a line between the nearest two recovered corners. The lost corner is reestablished at a proportionate distance on the true line connecting the recovered corners. Single proportionate measurement is often used when one or more section corners need to be established. As discussed above, this method is only employed when there is no other recourse. Consider Fig. 7.7 where monuments were found at points 1 and 2. The missing section corner is halfway between points 1 and 2. The surveyor recovered the monuments at points 1 and 2 and measured 160.90 chains between the two monuments. The distance along the line is calculated based on the exact proportion of the record distance. For example in Fig. 7.7, the total record length is 160.00 chains and the record distance to the boundary corner is 80.00 chains. The distance would therefore be based on the proportion:

$$Corner\,Distance = \frac{40.00}{80.00} * 160.90 = 80.45$$

In our example, the record location of the corner is exactly halfway along the line, so the remaining distance to the other corner would also be 80.45 chains.

7.12 Double Proportionate Measurement

Double proportionate measurement is used when four corners exist; two being on a meridional line and two being on a latitudinal line and the intersection of both lines is the corner. Note that the original bearings of the line are not used for locating the line. The final corner location is based solely on the proportional distances along the two lines. Consider Fig. 7.8 where the missing corner of four intersecting sections is missing. Monuments have been recovered north and south of the

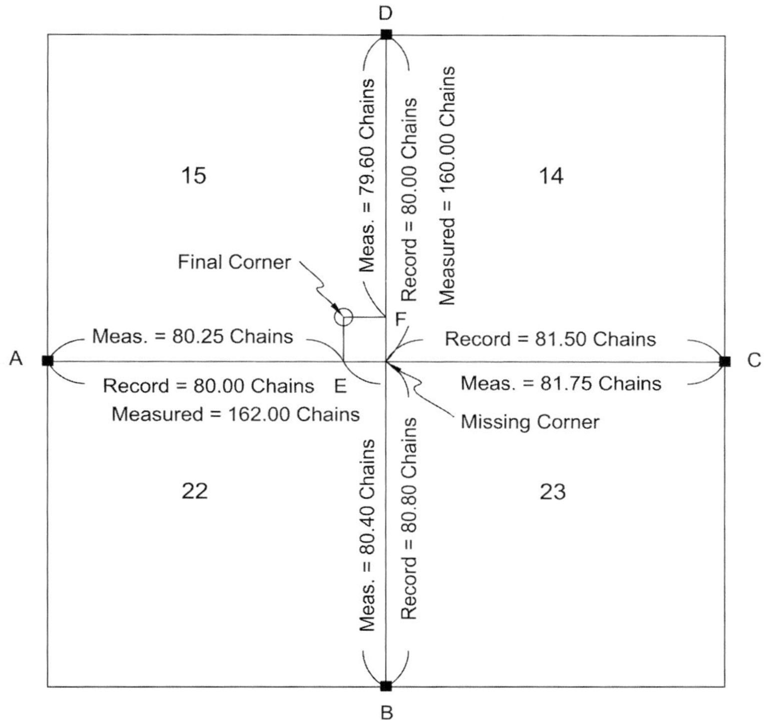

Fig. 7.8 Double proportionate measurement

latitudinal line. Monuments have also been recovered east and west of the meridional line. A temporary point will first be established at point E by using a proportion of the record distances to the measured distance between the monuments as follows:

$$Distance\,A - E = 162.00 * \frac{80.00}{80.00 + 81.50} = 80.25\,chains$$

The total measured length of the line A–C is 162 chains. The record distance of the north line of section 22 is 80.00 chains and the record distance of the north line of section 23 is 81.50 chains. As a check, we can calculate the distance from E to C as follows:

$$Distance\,E - C = 162.00 * \frac{81.50}{80.00 + 81.50} = 81.75\,chains$$

$$Total\,Distance = 80.25 + 81.75 = 162\,chains$$

Now that the distance to point E has been calculated a temporary mark would be placed on the line A–C at point E.

We would use the same procedure to calculate the distance to point F on the meridional line and place a temporary point at F on the line from B to D. Once points E and F have been marked on the ground, a line would be run true north from point E and true west from point F. The intersection of these two lines would mark the lost corner. In actual practice, using a total station or GPS and calculations based on coordinate geometry, the coordinates of the final corner would be calculated in the office or field computer and an angle and distance turned from a traverse point to the final corner. This would relieve the crew from setting temporary points E and F.

Chapter 8
Making a Boundary Survey

8.1 How Does a Boundary Survey Proceed?

All boundary surveys are made by following four basic steps. These are:

1. Legal research to gather record evidence.
2. Field work to gather physical evidence.
3. Calculations to adjust survey data and to reconcile record and physical evidence.
4. Field work to set property corners.

These steps will be discussed in detail in the following chapters. Sometimes, the steps may appear to merge or occur out of order, but they occur nonetheless. For example, a simple stakeout of an existing subdivision lot might be accomplished with only one trip to the field. The following is such a scenario. The client's deed calls for a plan which is subsequently found on record in the registry of deeds. This plan is then used to calculate the location of lot corners based on physical monuments shown on the plan. A trip to the field discloses that the monuments do exist and that they are in good agreement with each other. The surveyor then uses these monuments to set the corners of the client's lot.

The preceding is possible only when the record information and physical evidence are in good agreement. Although this does occur, in the real world the surveyor frequently encounters problems that must be resolved before the lot corners can be set.

P. Gay, *Practical Boundary Surveying*, DOI 10.1007/978-3-319-07158-9_8

8.2 Original Surveys and Retracement Surveys

Boundary surveys can be divided into two basic types: Original Boundary Surveys and Retracement Boundary Surveys, sometimes called resurveys. Combinations of both in one survey are possible.

Boundary surveys can be divided into two basic types: Original Boundary Surveys and Retracement Boundary Surveys, sometimes called resurveys.

An original survey is where a parcel of land is created for the first time. For example, Farmer Jones owns Willow Brook, a 100 acre ranch. Mr. Jones decides to convey a one acre parcel of land to his daughter so that she can build a house. The location of the proposed one acre parcel is near the middle of Mr. Jones ranch and it will not be bounding on land belonging to one of his abutters. In effect, after Mr. Jones conveys the parcel to his daughter, he will be the only abutter. Mr. Jones retains a surveyor, Mr. Perambulator, P.L.S, to establish the boundaries of the one acre parcel. Mr. Perambulator sets concrete monuments at each corner of the new lot and draws a plan showing the metes and bounds of the lot. This parcel represents an original boundary survey because the property lines created as a result of the survey and the subsequent conveyance to Mr. Jones' daughter will be new boundary lines—lines that did not exist before the survey. An original survey **creates** boundaries.

A retracement survey or **Resurvey** is a survey that attempts to duplicate an original survey. Let us reconsider the previous example. Suppose that Mr. Jones' daughter acquired the one acre parcel in 1945. In 1965, twenty years later, she decides to erect a fence along her boundaries. She wants to be sure she is erecting the fence wholly on her property so she searches for the concrete monuments set by the original surveyor but discovers they are missing. This is not an uncommon occurrence, particularly on a farm where the use of agricultural machinery over many years may damage or destroy boundary markers. As the original surveyor, Mr. Perambulator, has passed on, she retains another surveyor, Mr. Hightop, P.L.S. to set new monuments. In surveying the lot, it will be Mr. Hightop's duty to reestablish the boundaries at their original locations—the locations established by Mr. Perambulator in 1945. A resurvey does not establish new boundaries. It is sometimes said that in a retracement survey the new surveyor "follows in the footsteps of the original surveyor".

It is common for a survey to contain elements of an original survey and a retracement survey. In the previous example, suppose Mr. Jones' property, Willow Brook, was bounded on the west by an abutting property called Westlawn. Mr. Jones' daughter would like the one acre parcel to come out of Willow Brook where it abuts Westlawn. In effect, the west boundary of the new lot would be common boundary between Willow Brook and Westlawn. The survey of the common boundary would be a retracement survey because the property line separating Willow Brook from Westlawn is an original property line in existence at the time of the new survey. The survey of the remaining lines would constitute an original survey because these lines did not exist prior to the new survey.

Retracement surveys are very common. Boundary surveyors are routinely called upon to resurvey boundaries which may have been established many years ago by a surveyor who is unknown or little known to the present surveyor. Whenever a surveyor establishes boundaries, the surveyor assumes liability for the work. If the current surveyor accepts the work of the prior surveyor, he or she is liable for that work. It is therefore prudent for the current surveyor to extend the scope of the modern survey to include the evidence and research necessary to insure that the work of the original surveyor was correct and that the boundary locations so established are reasonable. For example, a client may have had a boundary survey of a large tract performed many years ago but the monuments for one or two corners have recently disappeared. The client wishes to have the missing markers replaced and he retains a new surveyor to reset the corners. In order for the current surveyor to replace the missing corners she will probably have to rely on certain remaining monuments set by the original surveyor. If the original monuments were set in error or if they have moved, the boundaries established by the current surveyor may be incorrect and she will be liable for damages. It is therefore necessary for the current surveyor to extend the new survey far enough so as to insure that the work of the original surveyor was performed correctly. In some cases, it does not require much additional work for the current surveyor to be comfortable with the original surveyor's work. In other cases, however, a complete resurvey of the entire parcel may be required. Perhaps understandably, clients sometimes think that because they have already paid a surveyor to survey their property, it is only necessary for the new surveyor to use the existing information to reestablish a corner or line. In their mind, reestablishment of the points should be easy and the cost for the new survey should be minimal. Unfortunately, this is not always the situation. In such cases the new surveyor must understand the client's position and explain why it is in the client's best interests to insure that the resurvey be performed comprehensively.

Chapter 9
Boundary Survey Research

Research is the first step in the performance of a boundary survey. A surveyor will not generally begin to have an appreciation of the level of difficulty of a particular survey until the research is well under way.

A surveyor will not generally begin to have an appreciation of the level of difficulty of a particular survey until the research is well under way.

Prior to beginning research, the surveyor has usually met or talked with the client regarding the client's requirements. Because it is the surveyor's duty to set boundary markers based on the record description, it is good practice for the surveyor to ask that the client furnish the legal description of the property that is to be surveyed. Sometimes clients do not have or cannot find a copy of their deed so the surveyor will need to obtain it from the recorder's office. Depending on the requirements of the particular boundary survey, the surveyor will, at a minimum, need to obtain copies of the deeds of all abutting properties. The surveyor will also need to get copies of any plans of nearby properties and any street layouts that affect the client's property. Before we discuss methods of legal research we need to have an understanding of the nature of real property and deeds.

9.1 The Concept of Real Property

A deed conveys title to an interest in property. In order to appreciate what this means, the reader must have an understanding of the nature of real property. In broad terms, the law deals with two basic types of property: real property and personal property.

P. Gay, *Practical Boundary Surveying*, DOI 10.1007/978-3-319-07158-9_9

Real property is land and generally whatever is erected, growing upon or affixed to the land. Real property consists not only of the land, but buildings, stone walls, fences, wells or driveways. Real property can consist of a right of way over abutting property, which is necessary to provide access to the subject property. In the case of a right of way, it is sometimes said that such things are **Appurtenant** to the land because they are incidental and necessary for the use and enjoyment of the land. Appurtenant means a right or restriction that goes with or attaches to a property.

Personal property is anything that is subject to ownership that is not real property. Personal property can be an automobile, television, watch, computer software, money, stocks or any tangible or intangible property subject to ownership, as long as it is not real property.

Title to real property consists of ownership of the elements constituting certain rights or interests in a particular parcel of land. The term **Fee** means an estate in land. It is the ownership interest in land which a person holds. It is possible to have title to land that is absolute and exclusive. This is the best title available and it is called **Fee Simple Absolute**. It is also possible to have lesser forms of title wherein other persons in addition to the property owner have certain rights to the property.

Rights or interests in real property may be compared to a bundle of sticks which the property owner holds in their hand. If the owner holds all of the sticks it is said they hold Fee Simple Absolute title to the land. Their ownership is exclusive because no other person has any of the sticks.

Suppose, however, that the property owner grants a right of way over his property to a neighbor, so the neighbor may have more convenient access to some portion of her property. In granting the right of way, the property owner has given away some of the sticks, he no longer has fee simple absolute.

A common occurrence is when a person borrows money from a lending institution for the purchase of property. The person purchases the property then gives the lending institution a mortgage on the property. In some jurisdictions a mortgage is legally construed as an interest in property so in those states the owner has given away some of the sticks to the lending institution.

Another example is where a person encloses an abutter's property with a fence and treats it as their own for a period of time sufficient for the property to be acquired through adverse possession. Here again the property owner has lost some of the sticks. In this case it is possible that the owner may not even realize they no longer have Fee Simple Absolute ownership.

9.2 Deeds

A Deed is a writing, signed by the grantor, conveying title from the grantor to the grantee. The **Grantor** is the person or persons conveying the property. The **Grantee** is the person or persons acquiring the property. Deeds must meet certain statutory and common law requirements regarding their construction. Only a few of these requirements concern the surveyor. The first four requirements have already been stated: a grantor, a grantee and, in compliance with the Statute of Frauds, a writing signed by the grantor.

A Deed is a writing, signed by the grantor, conveying title from the grantor to the grantee.

Because others can have rights in a parcel of land which may not be evident in the grantee's deed, a deed cannot be considered actual proof of ownership. A deed is, however, evidence of ownership.

9.3 The Statute of Frauds

In the year 1677 an English statute, entitled **The Statute of Frauds**, was enacted which required that no legal action for enforcement of certain contracts could be maintained unless there was a writing signed by the party to be charged. Conveyancing laws in this country are based on a provision of the Statute of Frauds which required that all interests in real property be evidenced by a writing to be enforceable.

Although there are exceptions, the modern day rule is, to be enforceable, conveyances of property or interests therein must be in the form of a written deed or other written instrument. The instrument must be signed by the party that is to be held accountable. Where the instrument is a deed, the accountable party is the grantor—the person who sold the property. Verbal agreements having to do with interests in real property, without more, are not enforceable. So, if your neighbor verbally offers to purchase your house and you verbally accept the offer, there is probably no enforceable agreement and both of you can walk away from the deal. There are exceptions to the rule such as the case involving partial performance: where the neighbor paid you some money, or made permanent improvements to your property.

In England prior to the Statute of Frauds, conveyances of interests in real property did not require written instruments such as a deed. Land was transferred by "feoffment" with "livery of seisin". The **feoffer** (grantor) and **feoffee** (grantee) met on or near the property and the feoffer announced the transfer in the presence of witnesses. The feoffer handed the feoffee a clump of earth or twigs as a symbol of the conveyance. Because there was no writing, the witnesses served as a means of public notice of the transfer. In Teutonic law a group of small boys from the area was assembled and required to observe the **Livery of Seisin**. Each was given a sharp hit on the head to impress solemnity of the occasion into their memory.

The requirement of a writing now exists in all American jurisdictions. The writing commonly used to evidence a conveyance is called a deed.

9.4 Deed Descriptions

A deed must identify the property being conveyed. The portion of the deed that describes the property is referred to as the **Legal Description**. This is the section of the deed that is of most interest to a surveyor. The description begins with some point of beginning such as "Beginning at a point in the westerly line of Main Street...", and usually ends with the area of the parcel, such as "Containing 1.45 acres more or less".

Most modern deed descriptions are what are called **Metes and Bounds** descriptions. These descriptions describe each boundary line in succession by giving a bearing, a distance or both for the lines. The **Metes** are the measurements: the bearings and distances. The **Bounds** are the monuments fixing the location of the line. A bound could be of the name of the abutter or a stone wall or a stone monument (stone bound). Figure 9.1 shows an example of a description drawn from a plan.

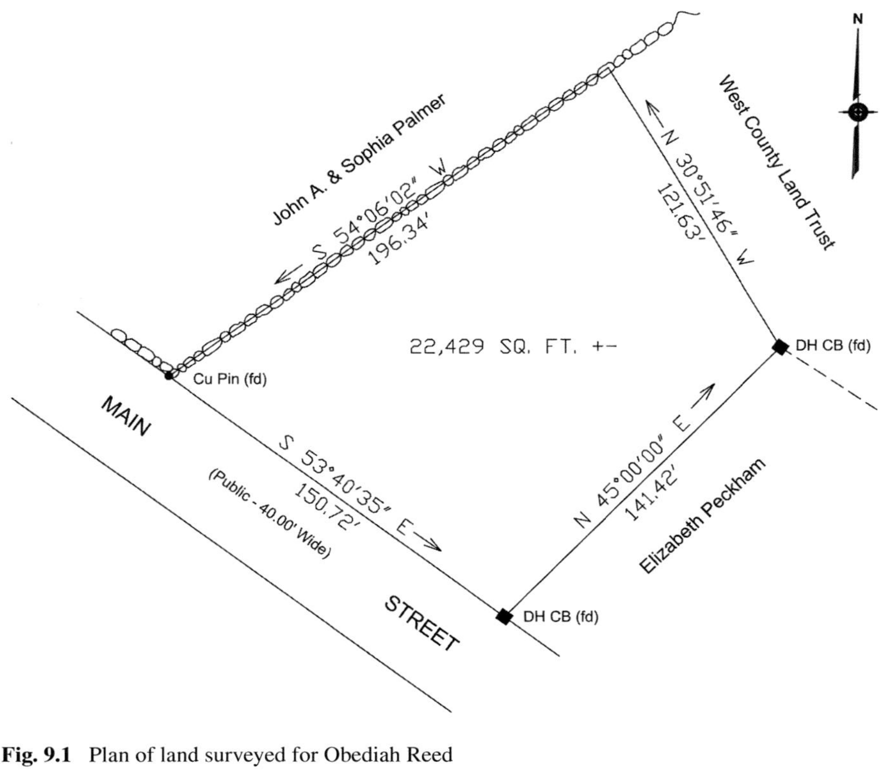

Fig. 9.1 Plan of land surveyed for Obediah Reed

Beginning at the southerly corner of the lot to be conveyed at drill hole set in a concrete bound in the northeasterly side of Main Street;

thence, North 45° 00′ 00″ East one hundred forty-one and 42/100 (141.42) feet by land belonging to Elizabeth Peckham to a drill hole set in a concrete bound and to land belonging to West County Land Trust;

thence, North 30° 51′ 46″ West one hundred twenty-one and 63/100 (121.63) feet by said last named land to a stone wall and to land belonging to John A. and Sophia Palmer;

thence, South 54° 06′ 02″ West one hundred ninety-six and 34/100 (196.34) feet by said wall and by said last named land to a copper pin set in the corner of the wall and to the northeasterly line of Main Street;

thence South 53° 40′ 35″ East one hundred fifty and 72/100 (150.72) feet by Main Street to the point of beginning;

Containing 22, 429 square feet, more or less.

Said land is shown and further described on a plan entitled "Plan of Land in Freeport, MA, Surveyed For Obediah Reed, January 12, 1943, Scale 1 inch = 40 feet" recorded in West County Registry of Deeds Plan Book 77, Page 34.

There is an old saying that a picture is worth a thousand words. The reader is invited to compare the plan shown in Fig. 9.1 with the above description and decide which is easier to interpret. If there is no plan available, it is usually necessary to draw a sketch from the legal description so that the property boundaries can be more easily visualized.

In some cases where a deed is created from a subdivision plan, the metes and bounds description is omitted and the lot is merely described by giving a lot number and a reference to the plan. The description might be written as follows:

A certain parcel of land located in Newburthy, Vermont described as lot 23 as shown on "Subdivision of Newburthy Farms, Newburthy, Vermont, Dated July 22, 1985, Scale 1″ = 50′, Macafee Engineers and Surveyors, Inc., Newburthy, Vermont", recorded in Plan Book 49, page 123.

Usually, when a subdivision lot is to be conveyed, the description contains both a metes and bounds description and a reference to the lot number shown on the subdivision plan.

In the western part of the U.S. where the sectionalized system of lands exists, descriptions often are in the form of township subdivisions. This is described in more detail in the chapter on the Public Land Survey System.

9.5 Deed Calls for Plans

Where a deed calls for a plan, the plan is **Incorporated** into the deed description. Incorporated means the document being incorporated is treated as though it were physically a part of the document which references it. When a document is incorporated into another document, the incorporated document has the same legal weight as the document referencing it. In the examples above, the descriptions referenced recorded plans. The plan must be obtained and interpreted as though it were a part of the deed. In construing the legal description of the property the plan will be given weight equal to or perhaps even greater than the words written in the deed. Often, plans will contain much more information than the deed. For example, a subdivision plan will commonly show a number of lots and streets whereas in a deed describing one of the lots, language in the description will be limited to the bearings and distances necessary to describe only the lot being conveyed. In order to establish the boundaries of a lot on the ground, surveyors will often need to go well beyond the deed description in order to find enough information to correctly locate the lot boundaries. Having the plan incorporated into the deed gives legal authority to the additional evidence shown on the plan.

Where a deed calls for a plan, the plan is Incorporated into the deed description.

9.6 Warranty and Quitclaim Deeds

Deeds are generally either **Warranty Deeds** or **Quitclaim Deeds**. When a grantor conveys property using a warranty deed, the grantor warrants or guarantees that the title to the property is free from certain defects. If it is later determined that title defects existed at the time of the conveyance the grantor would be liable in suit for damages.

In most jurisdictions a quitclaim deed conveys whatever interest the grantor has in the property. The grantor makes no promise to the grantee as to the quality of the title being conveyed. The grantor is merely conveying what he has, whatever that happens to be.

Although a grantee would generally prefer a warranty deed over a quitclaim deed, the advantage may not always be significant because, in order to collect damages, the grantee would probably have to sue the grantor who may be insolvent, or out of the jurisdiction. In addition, the legal costs of suing the grantor may be substantial.

9.7 The Recording System

Each state is responsible for maintaining records affecting title to land within the state. This system of records is referred to as the recording system. In many states, the recording system is maintained on a county level but in some states each town or city keeps the records. A handful of states also have a system called the Torrens System which registers title to land.

The recording system can be thought of as a "library" containing all of the documents or records relating to legal transactions affecting land within the jurisdiction of the system. Each of the documents is indexed by the government employees. These documents are placed chronologically into consecutively numbered volumes. The government does not take responsibility for the contents of the documents and it takes no position relative to the state of the title evidenced by the documents. The responsibility for placing conveyances on record is solely with the individual parties to the transaction. The government only records documents presented for recording. These documents must often meet certain formal requirements imposed by the jurisdiction in order to be accepted for recording.

9.8 Recording System Index

A recorder's office contains two basic types of volumes: **Indexes** and the **volumes containing copies of the recorded documents**. Two types of indexing are in common use. One is a grantor–grantee name index and the second is a tract or parcel index.

The first type indexes documents by the names of the parties. In this system there are two sets of indexes: the **Grantor index** and the **Grantee index**. As noted earlier, the grantor is the person who sells or transfers property and a grantee is the person who acquires the property. A grantor index volume will contain the names of grantors arranged in alphabetical order. Each index volume might cover a year or span several years of recorded documents, depending upon the capacity of the volume and the number of transactions. A typical index volume consists of an alphabetical listing of grantors who recorded documents between certain dates, such as during the past year.

If a researcher only has the name of a grantor and does not know the approximate year of the transfer it may be necessary to search through many volumes in order to find the deed reference. The same is true for the grantee index.

An example may make the preceding description of grantor-grantee indexes clearer. Suppose "Mary Conveyor" conveys a parcel of land to "Robert Byre" in January of 1965 and the deed is immediately presented for recording. The recorder's office will place an entry in the current grantor index listing Mary Conveyor as grantor. Depending on the policy of the particular recorder's office, additional information may also be placed in the record. This information is usually the name of the grantee, the date of the conveyance, the general location of the property and the book and page of the volume in which the deed can be found. The same information is also recorded in the grantee index but in this case entry would be indexed in the name of Robert Byre, the grantee.

The grantor–grantee index allows a person to find a conveyance if they know the name of either the grantor or the grantee. Using the preceding example, suppose you knew or suspected that Mary Conveyor purchased many parcels of real estate in 1941 and wanted to know if she had sold a particular parcel sometime between 1941 and the present. You could start with the most recent grantor index volume and look up the name "Conveyor, Mary". Say the current volume spanned 1999 to the present.

Finding no transfer in the current volume you would take the preceding grantor volume from the shelf which might span the years 1995–1998. You would repeat this procedure until you either found the transfer or until you reached the volume containing the year 1941, the year you knew Mary Conveyor purchased the property.

Here is another example. Suppose you have agreed to purchase a house and land from Robert Byre and want to obtain copies of the deeds of all of the prior conveyances of the subject parcel of land. You do not know from whom Robert Byre purchased the property or when he bought it. You would go the current grantee index and look under Robert Byre. Because Robert Byre is the current owner he would have been the grantee in the deed. In this case you would have to go back through several volumes until you got to the grantee volume dated 1965 before you discovered that Robert Byer purchased the property from Mary Conveyor in January of 1965. The record would give you the book and page where the deed was recorded.

To trace the title further back in time you would start with the date January 1965 and look in the grantee book for that year under Mary Conveyor. You would again look through a number of books, each one earlier in time, until you got to the 1941 book which listed Mary Conveyor as grantee of the subject parcel. This record would give the name of Mary Conveyor's grantor which you would then use as grantee to continue to trace the title back through the grantee books. The preceding is an example of what is called the **Chain of Title**. For any parcel of land that has gone through a sequence of conveyances each deed represents a link in the chain.

The second type of indexing system is the tract or parcel index. This approach is considered by many knowledgeable practitioners to be a much better system than the grantee–grantor index. Some of those familiar with the system question why it is used in such a small number of states. The tract index has a page or several pages of entries relating to some tract of land such as a quarter section, subdivision, or individual parcel of land. In a tract index it is the parcel of land that indexes the recorded documents associated with it. If one is interested in the chain of title of a particular parcel one goes to the index for that parcel and retrieves a history of recorded documents affecting title.

Most deeds contain a reference to the book and page of the deed under which the grantor acquired title. In some jurisdictions this reference is a requirement and a deed will not be accepted for recording unless it contains the necessary language.

9.9 Chain of Title Problems

From a surveyor's perspective, one purpose of searching the chain of title to property is to determine if there have been conveyances out of a parcel of land which could change the boundaries of the property being surveyed. Another reason that the chain of title is important is to determine junior and senior rights of parcels of land derived from a common grantor. We will discuss the importance of this later on, but suffice it to say that it is important, for a number of reasons, to be able to trace each deed in the chain.

> One purpose of searching the chain of title to property is to determine if there have been conveyances out of a parcel of land which could change the boundaries of the property being surveyed.

If a deed is recorded but is not recorded in the chain of title, or if a deed is not recorded, the deed is called a "wild deed" and in most jurisdictions the deed will be treated as though it were not recorded. As an example, consider the following. "Anthony" conveys Hidden Meadow to "Barbara" who fails to record the deed. Later Barbara conveys Hidden Meadow to "Claudette" who records the deed. If you attempted to trace Claudette's chain of title backwards you would not be able to do so because the deed from Anthony to Barbara was not recorded. The recorded deed from Barbara to Claudette would tell you that Barbara was the grantor but a search of the grantee index for Barbara as grantee would fail because the deed from Anthony to Barbara was not recorded. The deed from Anthony to Barbara would be a wild deed. A similar result would occur if a deed was recorded but improperly indexed.

In some jurisdictions, when a property owner dies and leaves property to an heir, title to the property passes immediately upon the death of the owner. Title to the property changes but there is no deed on record to evidence the transfer. In such cases it is necessary to search probate records.

9.10 Constructive Notice and the Recording System

Under the law, people are charged with notice of a prior conveyance as long as there is a deed on record, in the chain of title, describing the transfer. This is sometimes called **Constructive Notice** because a person may not have actual notice. For example, the person may not have bothered to check the recorder's office so they would not have actual notice. However, whether or not the person bothered to look up the deed is immaterial. The fact that the deed was on record means the person, indeed everyone in the world, has notice. So, either a person has actual notice or they are presumed to have notice. In either case the result is the same, constructive notice means that the law considers everyone to have notice, whether they actually have it or not. Think about it. Wouldn't the value of the recording system be substantially impaired if people could simply claim they didn't know about a recorded conveyance, and as a result they should not be held accountable?

Sometimes people forget or don't bother to record a deed. Suppose you purchase a vacant parcel of land and forget to record the deed. In many states, if the grantor subsequently conveys the same property to some innocent third party (a grantee who was not told and did not know that you already purchased the property), and the grantee records the deed before you record your deed, you will lose the property. The grantee would be protected by the recording system. She is presumed to be on constructive notice that there are no conveyances on record of the subject property. If she had researched the record, her research would have

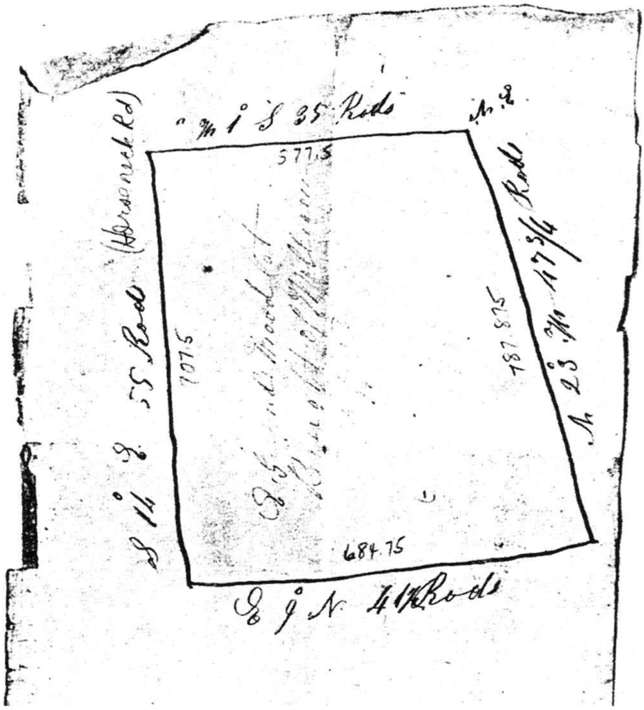

Fig. 9.2 An ancient handwritten plan held in private records. Notice that distances are in rods. Also notice that the bearing E9°S does not conform the modern system which would be written as S81°E. Someone has converted the rods to feet and written them in on the plan

indicated that the grantor still owned the property and the grantor would therefore have been in a position to lawfully convey it. As long as she did not actually know that you purchased the property first she would not be on notice of the prior conveyance. She therefore takes the property and you lose it.

The forgoing doesn't mean that deeds or plans that are not recorded have no value. An unrecorded deed is valid as between the seller and buyer. It is just not valid against innocent third parties, because these parties would not have notice of the document. Also, plans or deeds kept in private records should always be scrutinized because it is possible they could have been altered in some way (Fig. 9.2). Privately held documents may be admissible in a legal proceeding providing that some person is available to testify as to the validity of the documents.

9.11 The Torrens System of Title Registration

A small number of states have established a system of land records that is based on the **Torrens System**. The system is named after Richard Robert Torrens who adopted the English system for registration of ships to the registration of real property in Australia. It was noted earlier that the system of land records in common use does not concern itself with title. The government merely records documents that are presented to it by the public. In a title registration system the government certifies title to land. Once a parcel of land has been registered (not recorded as a deed would be recorded in a recorder's office) the state creates an original "certificate of title". The certificate of title is kept by the government and a duplicate certificate of title is issued to the property owner. The certificate of title may be recorded in the same recorder's office where other unregistered land titles are recorded, although it may be kept in a separate room or section. The certificate of title lists the names of the owners of the property and notes any and all encumbrances to the property. Encumbrances can be liens, mortgages, easements, leases, or other claims.

The certificate of title is conclusive proof of the current state of title of the subject property. In other words, the government is guaranteeing the state of title. From a researcher's point of view, doing a title search of registered land is as simple as looking up the certificate of title.

In some jurisdictions, such as Massachusetts, the authority responsible for registration, called the Land Court, prepares an official plan of the property that is filed along with the certificate of title. The actual survey of the property is made by a private surveyor retained by the property owner. The property owner is responsible for paying the surveyor but the surveyor's duty in establishing the property boundaries is owed to the Land Court. In this case the state is guaranteeing both the title and the metes and bounds description of the property. The foregoing is not to say that valid claims to the property may not exist. However, the government maintains a fund to pay damages, if it is determined that a valid claim to the property does exist.

Although the Torrens system has a few benefits, it has many drawbacks. For example, one may have to deal with an arcane and inflexible bureaucracy. The requirements imposed by the system may make the survey prohibitively expensive. In some systems, once land is registered it is difficult or impossible to un-register it so that all subsequent divisions of the land are subject to the time and expense of compliance. Where an ordinary survey can be completed and recorded in days or weeks, it may take years to get it through the Torrens system. This could easily kill a deal involving the sale or subdivision of the registered property. Many, if not most boundary surveys can be performed by a licensed surveyor. In contrast, when dealing with registered land it is likely that an attorney will also be required. This will probably increase the cost of the survey substantially. Finding an attorney and land surveyor who is familiar enough with the system to efficiently do the work may be difficult. Some surveyors view the difficulties in dealing with the system more trouble than it is worth and many do not want to take on the work. Some clients do not understand that a registered land survey can cost many times what a conventional survey might cost and they are reluctant to spend the extra money. For many, if not most people, the drawbacks of the Torrens system far outweigh its benefits.

Title registration is voluntary. The title registration system coexists with standard recording procedures. The only titles that are registered are those which are presented by the property owner to the title registration office.

As has already been noted, a disadvantage of title registration is the expense and time required to complete the initial registration. The registration process begins when a petition is filed by the property owner. The petition is much like an action to quiet title which might be filed in any court of competent jurisdiction. The government must examine the title of the property presented for registration, give notice to abutters and other interested parties and schedule a hearing or trial, as necessary, in order to quiet any claims against the property. The procedure can take from 6 months to several years before a certificate of title can be issued. Costs are dependent on whether or not there are title problems and whether there are boundary lines or other issues that are contested. A contested registration can cost thousands or even tens of thousands of dollars.

One advantage of the title registration system is that, at least in some jurisdictions, adverse possession and prescription does not run against registered land. In states such as Massachusetts where the metes and bounds description of the land is established by the Land Court, surveyors must be mindful when surveying land abutting registered land that the boundaries cannot be altered. The property lines established by the Land Court are fixed and not subject to relocation by the surveyor when interpretation of the evidence would suggest another location.

9.12 Getting Oriented

When beginning research it is often helpful to have an overall picture of the area to be surveyed. Some old deed descriptions can be difficult to locate or to interpret, particularly if they contain errors, so having an overall plan of the area can save time. If the survey is of a lot in a relatively recent subdivision, and the lot does not happen to be on a subdivision boundary, then the subdivision plan itself may provide all the information which the surveyor needs. In most other cases an overall plan of the area is often helpful.

There are a number of potential sources for such plans. One source may be the city, town or county Assessors or clerk's office. Because Assessors need some way of keeping track of who owns what, most Assessors offices have plans showing the division of ownership in their jurisdiction. The quality of these plans varies from one office to another. Because new subdivisions are created frequently, the plans have to be continuously updated. If the Assessors are understaffed or do not have the budget to maintain the plans, they may be out of date. A word of caution is in order here. All surveyors know or should know that Assessors plans are prepared for assessment purposes only. Because they do not generally result from an actual field survey they should not be relied upon for describing legal boundaries. Unless an Assessors plan is specifically called for in the deed, it should not be used as a basis for a legal description.

Sometimes aerial photographs are available for the area in question (Fig. 9.3). Some Assessors offices use large aerial photos as a background and draw parcels

Fig. 9.3 An aerial photograph

representing ownership on an overlay. Often, the local conservation office will have aerial photos. It can sometimes be difficult to obtain good copies of these plans, particularly from small town offices that do not have large format reproduction capabilities.

Aerial photographs can be very useful in that they show physical features existing at the time the photograph was taken. If, for example, a deed calls for a stone wall but no wall currently exists, an old aerial photo may show the location of the wall. There are companies who advertise on the Internet, whose business is providing historic aerial photos. Sometimes these photos are available for the early part of the twentieth century.

Other possible sources of information are old atlases, U.S. Geological Survey maps, old maps in the possession of the local historical society and other old maps that may be on record in the registry of deeds or the office of the town or city clerk.

Many municipalities have or are in the process of implementing Geographic Information Systems (GIS). These systems make use of a database containing information that is referenced to geographic coordinates. A GIS system can contain many types of information such as the locations of wetlands, property lines, buildings, or just about any attribute that can be linked to a geographic location. The same GIS system has the potential to be used by a number of local government agencies in concert. For example, the assessors could use it to store land ownership information, the conservation commission could use it to show wetlands, the planning board could use it to show new lots and subdivisions, the highway department could use it to show the location and extent of public and private streets, etc. The same caveat that applies to assessors plans applies to GIS information. Because the information may not come from actual field surveys made by licensed surveyors it should not be used for establishing boundaries.

9.13 Getting Deed References

Once the surveyor has the **Locus** (property being surveyed) deed, it will be necessary to obtain deeds for abutting properties, and in many cases deeds for property beyond the abutting property. When surveying a lot in a city, it is not unusual to need deeds for all of the lots along the street for the length of the city block and sometimes all of the deeds in the entire block may be required. The same approach may be required when working in older subdivisions, particularly when there are known problems.

In order to find these deeds the surveyor will need to go to the recorder's office and look in the correct book and page for each deed required. The assessor's office in the city, town or county in which the property is located is often times a good place to find the book and page references to deeds. In many jurisdictions, land records are kept at the county level, not the local level.

As already noted, many Assessors' offices have maps showing parcels of land in their jurisdiction. These parcels are usually identified with map and lot numbers. The assessors may keep cards or a computer file relating the map and lot number to the title information for the lot. This information often contains the book and page numbers of the deed for the current and possibly previous lot owners. In such cases the surveyor can note the book and page numbers and then go to the recorder's office and obtain copies of the deeds.

If the assessors do not have the book and page numbers it will be necessary to note the names of the property owners, and hopefully the date of the conveyance, and use the grantee index in the recorder's office to find the deed references. Needless to say, this method can take considerably longer than if the book and page numbers are available directly from the assessors.

Sometimes assessors have other information available such as subdivision plans that have previously been approved by the local planning board. Assessors, particularly in small towns, are often very knowledgeable about real property transactions, and issues related to local property, so if the researcher is having difficulty finding something it might be a good idea to ask for help.

9.14 A Deed May Not Describe Current Ownership

The experienced researcher is aware that an existing deed for a property may not describe the current state of ownership. There may have been parcels sold out of the original deed which are not exempted from the original description. For example, suppose that A owns Single-A-Ranch, a 100 acre parcel. Over the years A sells off a number of lots. Because A's deed is already recorded and cannot be changed, the deed no longer describes the current state of ownership. A researcher surveying A's property always has to look in the grantor index to determine if A has conveyed some part of the original parcel. If the researcher determines that A has made conveyances, he will need to get copies of those deeds.

The same rule applies to acquisitions by A. If the surveyor is charged with surveying all of A's property she must search in the grantee index under A's name to determine if A has gained additional property.

9.15 Researching Plans

At some point during the research process the surveyor will need to determine if there are any plans on record or elsewhere that have the potential to affect the subject property, or provide information that may help to locate evidence of boundaries. Whereas deeds are almost exclusively found in the recorder's office, plans have a number of possible sources.

Sometimes a deed will reference a recorded plan. If so, the plan should be available in the same recorder's office as where the deed was found. Experienced surveyors know that just because a deed does not call for a plan does not mean that plans do not exist on record. Recorders' offices usually have a separate index for plans. Just how plans are indexed varies from one jurisdiction to another. Sometimes they are indexed by street names, sometimes by the name appearing in the plan title and sometimes by both. Street indexes are useful because it is often possible to find plans showing surveys of nearby parcels.

Most property in populated areas has some frontage on a public or private road or street. If this is the case, then the surveyor will need to establish the street line. Many public streets have been "laid out" by the city or town. The layout of a street usually consists of a plan showing directions and distances of street lines and the location of monuments set to fix the street lines (Fig. 9.4). Sometimes these layouts can be found in the recorder's office. It is not unusual for these layouts not to be recorded. In such cases the surveyor will have to go to the local highway department or department of public works in order to obtain copies of the plans.

Many highways are created and maintained by state agencies. Where this is the case, the surveyor will need to contact the state highway department in order to procure copies of the layouts. A similar situation may exist when a railroad forms a boundary. If the plans are not found in the recorder's office it will be necessary to contact the railroad engineering department to obtain copies of the plans.

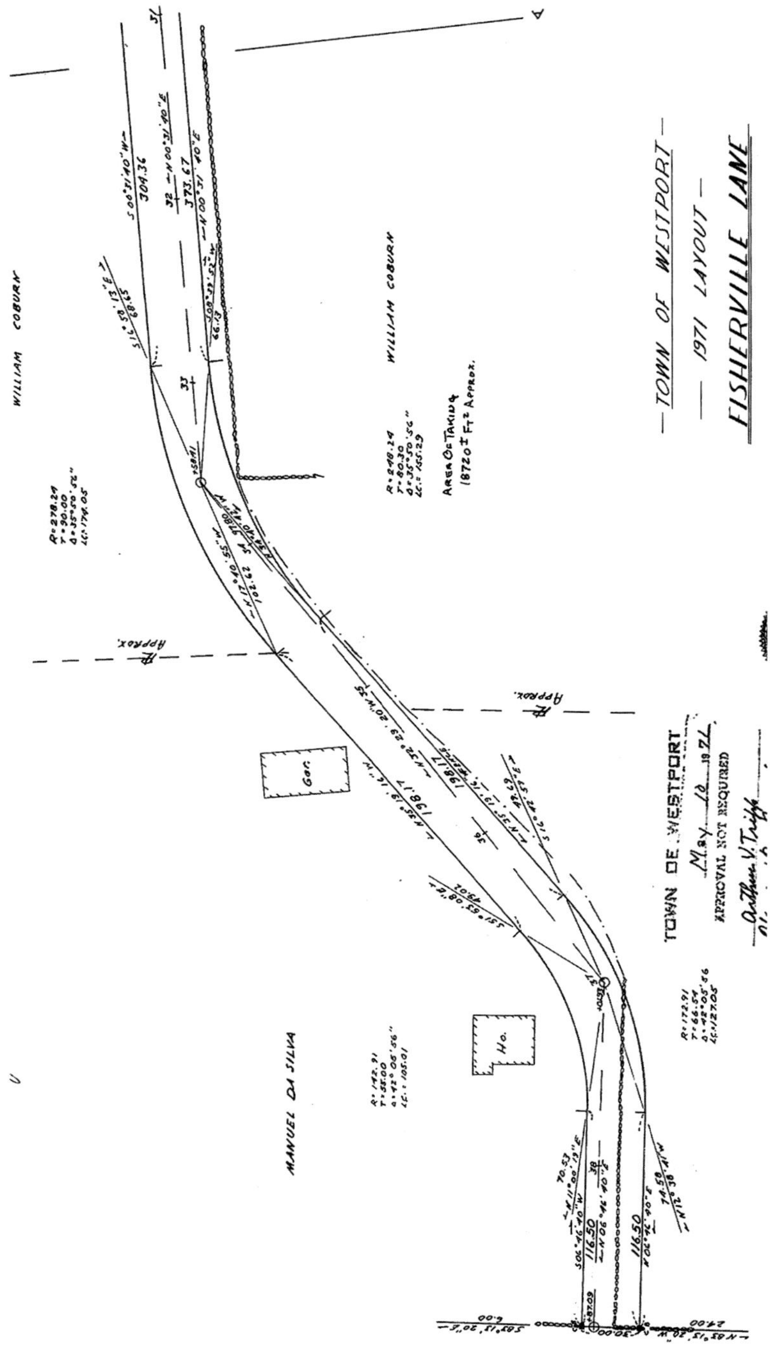

Fig. 9.4 A street layout made by a municipality

It is not uncommon for a surveyor to draw a plan or sketch of a surveyed parcel for the use of the owner with no intention of recording the plan. If the owner has the plan, it can be supplied to the surveyor. Often, the owner no longer has the plan or the survey was made for a previous owner now deceased. The only alternative the surveyor has in such cases is to contact other surveyors in the area and ask if they have any records pertaining to the subject parcel.

Abutting property owners often have unrecorded plans showing results of previous surveys. Sometimes these plans are the only available evidence of past surveys. A surveyor is well advised to make every attempt to inquire of abutters if they have any information that relates to their property lines. In the author's experience, most abutters would like to avoid the possibility of a surveyor placing boundary markers at a location unfavorable to them, so if they do have information regarding their claimed boundary lines, they usually want to bring it to the attention of the surveyor.

9.16 Problems Researching Plans

It should be clear from the foregoing that finding plans is not always an easy process. It often requires considerable imagination, perseverance and leg work on the part of the researcher to uncover the information.

One problem which often faces the researcher is the difficulty in getting plan copies. Plans are often quite large documents sometimes 24″ by 36″ or larger so they are not conveniently copied on standard copying machines, although they can sometimes be copied in small sections and glued or taped together later. In many recorder's offices some originals are so old and delicate that researchers are not allowed to place the plan books on the copy machine. The plan books are often very heavy and it is easy to damage the original plans. Sometimes the researcher has no choice but to use tracing paper and trace the original plans. Needless to say this is a very time consuming process and there is a possibility of copying a bearing or distance incorrectly.

Some recorder's offices keep plans on microfilm. Copies made from the film are often not to scale and are sometimes illegible. In many cases, even though the plans exist on microfilm or in reduced copies in plan books, the recorder's office keeps the original plans in an archive and they are available to be used for making copies. Some recorder's offices may discourage the researcher from requesting copies made from the original because of the extra work and expense in retrieving and copying. In such cases the researcher must keep in mind that the surveyor has an ethical duty to the client and the public to diligently search for and retrieve any information that may affect the location of the property lines. The surveyor must tactfully insist on being provided legible copies. In addition to understanding the technical and legal aspects of boundary surveying, the surveyor is sometimes called upon to exercise great patience and powers of persuasion. One purpose of the recorder's office is to make record information available to the public. If the provided information is not legible or is otherwise deficient then the office is not fulfilling its mandate.

Small battery powered scanners are available which can be placed on or dragged over a page. These devices store the image of the plan or other document in computer memory. When the researcher returns to the office the image can be sent to a computer and printed or printed using an infrared beam if the printer is so enabled. Another possibility is the use of a digital camera to capture images of a plan. The camera must have relatively high resolution and the ability to focus in such a way as to produce an acceptable image. In taking multiple images of the same plan, the camera must be held at exactly the same distance from the page or the scale of each image will be different, making it difficult to paste the images together. Although time consuming, images can be imported into a CAD program and enlarged or reduced to fit the desired scale.

9.17 Internet Searches of Land Records

Many of the research issues described in the preceding paragraphs are becoming less of a concern as recording offices are becoming computerized with internet access. Where such internet access exists, acquiring copies may be as easy as visiting the recorder's office web site, performing a search and downloading the document. In some jurisdictions documents can be downloaded and printed at no cost. In other cases, the documents can be viewed but only printed after payment of a fee. Many recorders offices are still in the process of scanning documents into their database. It is common for the latest documents to be scanned first and older documents added to the database as time and funds permit. A few searches by the author at the time of publication of this book showed that in some cases, online records only went back to the 1970s. For any documents older than this, it would be necessary to visit the recorder's office and obtain the records using a copier or other means. Because recorders' offices are often a part of the county government, the online availability of documents can vary substantially with the jurisdiction.

9.18 Mortgage Plot Plans

Sometimes, a client will mention to the surveyor that they have a "plot plan" showing their property. Often, these plot plans are what are commonly called **Mortgage Plot Plans**. These are sketches required to be made by lending institutions prior to lending money on a particular property. Their purpose is presumably to determine whether there are apparent encroachments and whether it appears that all of the buildings are actually on the subject property. Mortgage plot plans are usually not the result of an instrument survey. Often, measurements are made by a single person with nothing more than a tape. Lending institutions may place cost restrictions on the preparer who provides the plans so there is pressure to complete the work in as short a time as possible. Because of the limited amount of time that the preparer can expend on one of these plans, they usually contain disclaimers such as a note that the property is not the result of an instrument survey.

Lending institutions generally make arrangements to have these plans prepared and the borrower is charged a fee for the plans as a part of closing costs. Because these plans do not represent actual surveys of the property, some surveyors consider them to be a disservice to the property owner because many property owners believe that the plans represent the results of a survey of their property and as such, the plan can be relied upon. It could be argued that the owner's position, in this regard, is not wholly without merit, particularly when the owner paid a fee for having the plan prepared. In any event, these plans have little actual value to either a land surveyor or a property owner and should be accorded little to no weight as reliable evidence of property boundary locations, or even of the relationship between boundaries and structures.

9.19 Compiling the Research

As the process of gathering research progresses it will be necessary to compile the research. If there are many deeds and or plans, the surveyor may find it convenient to take the pertinent information from the individual documents and compile it in the form of a sketch. The sketch can then be carried into the field thereby reducing the number of documents that the surveyor will need to deal with. In the field, the surveyor must contend with the terrain and the elements and carry equipment. When the survey crew finds itself knee deep in the middle of a swamp besieged by mosquitoes, snakes and killer bees, it is usually not the most convenient time or place to try to read and make sense out of the archaic language of a dozen deed copies, perhaps handwritten in the 1800s and barely legible due to a worn out and under-funded copier in the recorder's office.

When the survey crew finds itself knee deep in the middle of a swamp besieged by mosquitoes, snakes and killer bees, it is usually not the most convenient time or place to try to read and make sense out of the archaic language of a dozen deed copies, perhaps handwritten in the 1800s and barely legible due to a worn out and under-funded copier in the recorder's office.

The information that the surveyor initially needs in the field will depend on the particular survey but will probably consist of the dimensions relating to property lines and the type and location of monuments called for in the record. If this information is clearly shown on the sketch the field work is more likely to proceed efficiently.

Chapter 10
Starting Field Work

Once the research is complete, or nearly so, the surveyor's next step is to go into the field to find monuments, to run survey lines, to locate evidence, and finally, to set lot corners. Sometimes, as the field work progresses, it will become apparent that more research is required. If the survey is a retracement survey, the surveyor's first order of business is to find the monuments called for in the record information. Even if the survey is to be an original survey, the surveyor will need to find some original monuments such as the street or highway bounds of the road which the property abuts.

For many surveyors, field work is one of the more enjoyable phases of performing a boundary survey. It rarely fails to be interesting. One never quite knows what to expect when in the field. Sometimes the weather is pleasant and other times it is rather unpleasant. Surveyors are routinely attacked by bees, gnats, mosquitos, ticks and, when property lines are contentious, sometimes their own kind. When working along roads and highways, vehicles are a nuisance at best and a hazard at worst. Working on farms and ranches, surveyors often have to contend with domestic animals (Fig. 10.1). Cows and horses are curious creatures and are attracted to instrument and prism tripods, often using them for scratching posts. A 1,000 pound animal has little difficulty tipping over a $25,000 robotic total station. This is likely to turn the surveyor's total station into an expensive mantel piece—a constant reminder never to leave the instrument unattended. Where large animals are free to roam in the vicinity of the survey crew, the person running the instrument must remain at their post at all times. Having equipment insured is sometimes a prudent investment.

Boundary surveyors rarely perform field work alone. Many surveys will require traversing, so three tripods with two prisms and a total station will be the minimum equipment required. Carrying all of this equipment around would waste too much time if a single person were to attempt it. There is also quite a bit of other equipment to be carried such as a prism pole, machete, hammer or hatchet, wooden stakes and other markers. Normally, a crew will consist of two or three persons, and sometimes

P. Gay, *Practical Boundary Surveying*, DOI 10.1007/978-3-319-07158-9_10

Fig. 10.1 Free range hogs in the field being surveyed

four. The person in charge of the crew is called the **Party Chief** or **Crew Chief**. The person running the instrument is called the **Instrument Man** or **Instrument Person**. The person walking around locating evidence is called the **Rod Man** or **Rod Person**. The traditional hierarchy of the positions begins with the party chief, who is the boss of the crew, then the transit person, then the rod person. The crew chief is usually the most experienced person in the crew. She is responsible for conducting the survey and having the final say regarding what needs to be accomplished and how it will be accomplished. The rod person is often the newest person on the team and often the least experienced. With a two person crew, the party chief is usually required to additionally perform the work of the rod person. Sometimes, if the instrument person is experienced, the party chief and instrument person will take turns running the instrument and locating evidence. When locating evidence, it is crucial that the person making the locations be very experienced in interpreting the nature of the evidence being located. This is important because the evidence will have legal significance in determining boundary locations. In such cases, the crew chief must either locate the evidence or at least observe the evidence and provide directions on how the evidence is to be located and noted in the record. It is beneficial that the crew operate as a team, with each member supporting the others in order to efficiently complete the tasks at hand. Although there should be a clear chain of command and clearly designated duties, this should enhance, not detract from the efficiency of the crew. Each member should be ready and willing to take on the duties of another crew member should it become necessary or desirable to do so.

10.1 Preparing for the Unexpected

When working in the field, surveyors must always be prepared for the unexpected. The old hand dug well Fig. 10.2 is flush with the ground surface. If the well were covered over with brush or snow it would be difficult or impossible to see before accidentally falling into it.

When working in the field, surveyors must always be prepared for the unexpected.

Sometimes land surveyors are fortunate enough to have the opportunity to work in remote parts of the country such as southeast Alaska, a location where brown bears outnumber people (Fig. 10.3). In many of these locations surveying equipment and supplies must be flown in on a float plane or brought to the location in a boat (Fig. 10.4).

Fig. 10.2 An old hand dug well bordered with field stones, a potential hazard to an unsuspecting person walking through the area

Fig. 10.3 Enjoying the sunset after a day of surveying in southeast Alaska

It may be necessary to move survey equipment and supplies with whatever means are available, such as the All-Terrain-Vehicle (ATV) moving an improvised "boat trailer" loaded with survey gear shown in Fig. 10.5. The robotic total station, in its yellow carrying case, can be seen in the rear compartment of the boat trailer which is precariously listing to starboard.

Figure 10.6 shows the beginning of the boundary survey with the total station set up on the edge of a gravel way and the prism tripod just visible on the left of the gravel way up the hill. A little further along the way the surface became so muddy and rutted that the ATV was barely able to make it through.

In remote areas the surveyor must deal with the logistics of ensuring adequate supplies and all of the equipment, including back up equipment that will be needed to perform the survey. When equipment has to be flown in at great expense, you don't want to be the one who forgot to pack the battery for the total station. The weather, such as the almost constant rain and drizzle in southeast Alaska can make the work difficult. High latitudes have very limited daylight in winter. In the location illustrated, the island was covered with heavy canopy of mature Sitka Spruce trees which prevented much of the available light from reaching the ground. In winter, on a cloudy day, the light can be so dim that it is difficult to see the target through the instrument, even at noon. In environments such as this a tracking total station can be a great asset (Fig. 10.7).

Surveys in such locations can present different sorts of challenges than those in more populated areas. The surveying technology and legal issues do not

Fig. 10.4 Flying people and equipment in a DeHavilland Beaver float plane into a remote location in southeast Alaska to perform a boundary survey

change appreciably but the environment, weather and logistics can place the work in an entirely different perspective. This can make working in such places very rewarding.

In this particular survey, the property owner had survived a confrontation with two brown bears a few weeks before. He offered to accompany the survey crew with his 375 H&H magnum rifle. His generous offer was accepted without hesitation. The bears must have been aware of this as they remained out of sight for the duration of the survey.

Fig. 10.5 ATV moving a "boat trailer" loaded with survey gear

Fig. 10.6 Total station and prism tripod on remote Alaska island

Fig. 10.7 Total station set up under sitka spruce canopy in the middle of the day. It can be so dark under the tree canopy that even with the excellent optics of the telescope it can be difficult to see the target

10.2 Dangerous Wildlife

Although habitat loss continues in the U.S., wildlife still exists in many locations where a survey must be performed. In some cases, increasing encroachments by humans into wild areas, with the resulting loss of habitat, may bring potentially dangerous creatures into populated areas. Habituation of wild creatures, for example, by leaving garbage or food scraps insecure where bears can get at them, increases the probability of confrontations with humans. This is likely to lead to unfortunate consequences for both the humans and the wild creatures. Surveyors spend a large part of their time outdoors. Having an intimate knowledge of the wildlife that might be encountered is an important part of the job. It is a subject not usually taught in courses on land surveying, but if you ask an experienced surveyor about it, you will be likely to hear about encounters with creatures ranging from bees to bears. When new highways are laid out, new subdivisions created, or new parcels are severed from larger parcels, surveyors are the first ones to go out into the field. This sometimes places them in habitats that have been largely undisturbed for years. In such cases, the survey crew is exposed to many species of wildlife. Sometimes a boundary corner may be located in a place that provides the perfect habitat for a poisonous snake, bees, poison ivy or other potentially harmful creatures or plants. If the boundary corner must be located or set, a surveyor may have no way to avoid the location and must walk to the exact spot. A prudent surveyor will be educated on the flora and fauna of the area in which he or she works. This knowledge will allow the crew to be prepared for any eventuality.

Only a few species have the potential to pose a threat to a survey crew. Indeed, humans are a much greater threat to wildlife than the other way around. The information presented here on dangerous wildlife is limited in scope and it is intended solely as an introduction to a few of the wildlife types a surveyor may encounter. Anyone entering an area where dangerous wildlife may exist should seek expert advice and familiarize themselves with the habits of the creatures, means of avoiding contact and what to do, should contact and/or injury occur.

Worldwide, the greatest wildlife threat to humans is the mosquito, because of its ability to transmit malaria. The World Health Organization estimated that in 2010, malaria caused 219 million clinical episodes and 660,000 deaths worldwide. In spite of malaria having been *eliminated* in the U.S., about 1,500 cases are reported each year, largely from people returning from travel abroad. Bee and wasp stings are in second place, mainly because many people have an allergic reaction when stung. Bees account for about 50 deaths per year in the U.S. Although not a wild animal, dog bites are high on the list, with about 200 deaths per year. Domestic dogs are found in urban, suburban and rural locations and surveyors should always approach unknown dogs with caution, particularly when they are loose and display aggressive behavior.

Worldwide, the greatest wildlife threat to humans is the mosquito, because of its ability to transmit malaria.

Lyme disease and Rocky Mountain spotted fever are carried by ticks. Lyme disease is carried by the Deer Tick and it is geographically concentrated in the Northeast with 95 % of cases occurring from New Hampshire to Virginia (Fig. 10.9). Although the number of fatalities may be less than 10 per year, reported cases exceed 30,000 per year. Deaths may be rare but the disease can be very debilitating, in some cases with long term effects. Rocky Mountain

Fig. 10.8 Bears are very common in some remote locations so surveyors should exercise caution and take steps necessary to prevent injury or death

spotted fever cases occur in most U.S. states, however approximately 60 % of the cases occur in five states: North Carolina, Oklahoma, Arkansas, Tennessee and Missouri. The disease can be severe and occasionally fatal if not treated within the first few days. Surveyors working in the field are routinely exposed to ticks and, in summer, it more likely than not that ticks will be found crawling on pants and shirts. The author recalls starting a survey on a warm spring day and finding more than 100 ticks in one morning.

Venomous snakes are found in most states in the United States. The Center for Disease Control and Prevention (CDC) estimates that 7,000 to 8,000 people are

Fig. 10.9 Deer Tick which transmits Lyme Disease

bitten by poisonous snakes each year, resulting in about 5 fatalities. The low ratio of fatalities to bites would increase substantially if people did not seek immediate medical care. People who work outdoors, such as land surveyors, are the most likely segment of our population to be bitten, so they should familiarize themselves with the species found locally, their habitat, what precautions should be taken and what to do if bitten.

About 25 species of rattlesnakes occur in the U.S. or Mexico. The most common species is the timber rattler, found in the eastern U.S. and the western diamondback, found in the western U.S (Fig. 10.10). They are the largest of the venomous snakes (more than 2 meters in length) and can accurately strike at up to one-third of their body length. They are named after rattles on the tip of the tail which make noise when the snake is threatened. The snakes are shy and tend to avoid humans but will strike if stepped on or provoked. Rattlesnake bites are excruciatingly painful.

Copperhead snakes occur in the eastern states as far west as Texas. They can be 18–36 inches in length. They are not aggressive, but when stepped on

Fig. 10.10 Western Rattlesnake Courtesy LA Dawson

they will strike. Copperheads are found around rocks and in wooded areas, often close to human habitation. They are well camouflaged and can be difficult to see.

Cottonmouth snakes or Water Moccasins are found in the southeast U.S. They are about 2 to 4 feet in length with coloration that is highly variable. Coloration ranges from dark crossbands over yellow to completely brown or black. They are semi-aquatic, living in swamps, lakes, rivers and ditches—all areas which surveyors are likely to frequent. On sunny days, cottonmouths can be found basking on logs or rocks near the water's edge. They tend to stand their ground and will bite when approached too closely. The snake got its name because when it opens its mouth, the white lining is exposed.

Suggestions for avoiding poisonous snakes, particularly rattlesnakes, include staying away from tall grass and piles of leaves, avoiding climbing on rock piles, wearing tall boots and wearing heavy leather gloves. Land surveyors routinely work in tall grass and often need to climb over rock piles, so they are uniquely susceptible to snake encounters. Traditional first aid treatment formerly suggested cutting the bite and sucking out the venom. The CDC does not recommend this practice as it presumably has a limited ability to remove an appreciable amount of venom. Instead, immediate medical attention is recommended where the appropriate anti-venom can be administered. Getting the injured person to an emergency medical facility in the shortest possible time substantially enhances the likelihood of a successful recovery.

> Suggestions for avoiding poisonous snakes, particularly rattlesnakes, include staying away from tall grass and piles of leaves, avoiding climbing on rock piles, wearing tall boots and wearing heavy leather gloves.

When a snake bite has occurred, the color and shape of the snake should be noted so the treating physician will know what type of anti-venom will be required. Many people now carry smart phones with cameras so it may be possible for a survey crew member to photograph the snake—keeping at a safe distance. The CDC does not recommend picking up or trapping the snake because of the chance of further injury or injury to another person, however others suggest killing the snake and bringing it in for identification. Many snake bites are "dry bites", where the snake does not inject venom. The frequency of dry bites varies by species but they may occur from 25 to 50 % of the time. Any snake bite victim should be rushed to an emergency medical facility without delay, whether or not the bite is believed to be a dry bite. Clearly, this is not the time to ask the victim to wait in the truck so the rest of the crew can spend a few more minutes locating that last monument.

Large predators, such as bears and mountain lions, are only responsible for one or two deaths per year on average, and encounters are extremely rare. Perhaps the helplessness of humans when attacked by a bear aggravates the perception of danger. The two species of bears found in the U.S. are brown bears and black bears. Brown bears are divided into two sub species, grizzly bears and Kodiak bears. Bears are omnivores, meaning they can subsist on plants or animals. Although they are sometimes perceived as being aggressive, bears will usually try to avoid contact with humans. When working in bear country, some experts suggest making plenty of noise so that bears are forewarned of human presence and have time to leave the area.

Bear attacks are usually defensive. Confrontations with bears often result from people not understanding bear behavior and acting inappropriately when in the presence of the animals. Young males, females with cubs and bears habituated to humans have the greatest potential to be dangerous. Even if a brown bear does charge, most of the charges are "bluff charges" which stop short of actual contact.

The Alaska Department of Natural Resources offers strategies to follow in bear encounters. So does Montana Fish, Wildlife and Parks. They suggest that if the bear is far away, you should go back or circle around and not disturb the bear. If a bear does see you, wave your arms and speak loud and in a low voice and slowly back away. Do not make eye contact with the bear. Do not run away. Brown bears can run 40 MPH and easily outrun the fittest of athletes. Running is also likely to trigger an instinctive predator-prey response causing the bear to chase the runner. If a bear does attack, curl up in a ball with your hands behind your back and play dead or lie face down with your hands behind your head. The fetal position is claimed to protect vital organs. The advice suggests that the bear will *usually* stop attacking once it realizes that the victim is no longer a threat. If the bear is a black bear, some experts advise not to play dead but to fight with whatever means are available.

The use of bear pepper spray has been recommended for people who frequent bear country, some claiming that, statistically, it is substantially more effective in preventing a mauling than firearms. It is important to only select pepper spray registered with the EPA for use on bears. If pepper spray is carried, it must not be kept in a backpack or other inaccessible location. It must be in a hip or chest holster where it is immediately available for use. The best sprays are claimed to have a range of about 30 feet. One downside to using pepper spray is that the wind may be blowing from the bear toward the sprayer, and spraying into the wind might cause the sprayer, not the bear, to become incapacitated. This would be doubly unfortunate for the victim.

One downside to using pepper spray is that the wind may be blowing from the bear toward the sprayer, and spraying into the wind might cause the sprayer, not the bear, to become incapacitated.

There have been an increased number of sightings of mountain lions in populated areas, primarily in the west and on the west coast (Fig. 10.11). Nevertheless, mountain lion attacks on humans are quite rare. The animals are carnivores and feed primarily on deer, helping to keep populations under control. Mountain lions will also kill domestic animals and pets. A mountain lion is capable of killing prey much larger than itself. The increased sightings may be evidence that, in future, it will be more likely for surveyors working in these areas to encounter mountain lions.

Unlike bear attacks, which are usually defensive, when a mountain lion attacks a human, it is probably looking for a meal. Mountain lions depend on stealth to get close to a victim before charging. They can sprint 50 MPH and jump 40 feet. They tend to attack from hiding places such as tall grass or embankments running alongside a trail. Once the prey is seized, a mountain lion will drag the prey to a place of safety before eating it. Experts say that attacks are more likely when a person is alone rather than in a group, although several recently documented attacks seem to have taken place with more than one person in the immediate vicinity.

Attacks tend to occur at dawn and dusk. People are advised never to turn their back on a mountain lion, although the creatures are so stealthy that it seems likely a victim will not even be aware of the predator until it is already on their back. It is also recommended that if attacked, the person should not run, they should

Fig. 10.11 Mountain Lion (U.S. Dept. of Agriculture)

fight back instead. There is some question as to whether pepper spray is effective on mountain lions, but it would seem to be worth carrying when working in areas where mountain lions are known to frequent, particularly if a firearm is not carried. It is important to recognize that even though the average person may be about the same weight as a mountain lion, he or she will be no physical match for the creature without a weapon of some sort. As predators, mountain lions survive because they are adept at killing. It would therefore be prudent to regard them as likely to succeed when opportunity presents itself.

People are advised never to turn their back on a mountain lion, although the creatures are so stealthy that it seems likely a victim will not even be aware of the predator until it is already on their back.

Some surveyors carry firearms when working in locations where there is a potential danger from wildlife. Because surveying involves carrying quite a lot of equipment, carrying a rifle or shotgun is one more, heavy, piece of equipment that must be carried. As one never knows when an attack might occur, to be effective, the weapon must be on the person, or within reach, at all times. Because of the weight and size disadvantages of rifles, handguns are sometimes preferred. For large animals, such as bears, a handgun capable of effectively dispatching the creature will have significant recoil and muzzle blast. Even the most powerful handgun will not be as effective as a large caliber rifle. In these cases, large caliber handguns may only be marginally useful, and even then, only in the hands of a highly experienced and expert marksman. Although a magnum handgun may be the only weapon available, a large bore rifle will certainly provide a greater margin of safety, with less chance of inadvertently wounding the animal.

When dealing with a charging brown bear, some experts believe the use of a firearm may be more dangerous than not using one. As most charges are false charges, use of a gun may unnecessarily kill or wound a bear that would not have done harm. During a close range charge, the shooter may only have time for a single shot. Wounding a bear is likely to aggravate the situation, making the bear even more intent on neutralizing the threat. Trying to stop a bear charge has been compared to trying to stop a small car heading toward you at 35 MPH. This is not to say that firearms should not be relied upon to save a person from an attack from a bear. However, the person using the firearm must be experienced and cool headed under pressure. Some claim that, with an inexperienced shooter dealing with such a large and dangerous predator, a firearm may provide a false sense of security, and do more harm than good.

Trying to stop a bear charge has been compared to trying to stop a small car heading toward you at 35 MPH.

The use of firearms, by the average person, is likely to be more effective when dealing with an 80–200 pound mountain lion as compared to a 1,000 pound bear. There is one report of a person being attacked by a mountain lion while a companion tried to beat the creature off the victim with a stick. The victim suffered substantial injury before the mountain lion finally decided to leave. In another case, when a person was attacked by a mountain lion, two companions threw stones at it until it finally released the victim and ran off. The victim was severely injured but survived. In both of these instances, authorities subsequently tracked the mountain lions and killed them. In both circumstances, the people involved did not expect an attack and so were not prepared to defend themselves. This is precisely why the stealth of a mountain lion is so effective, because it catches its prey off guard. In cases like these, an ordinary 38–45 caliber self-defense handgun would clearly be preferable to sticks and stones, particularly when used at point blank range. As it is probable the offending creature will be hunted down and killed anyway, every effort should be made to kill an attacking mountain lion as quickly and painlessly as possible in order to minimize injury to the victim and to eliminate the possibility of injury to another potential victim who may inadvertently encounter the lion. Unlike a brown bear's defensive tactics and the chance it will stop the attack once it believes there is no longer a threat; a mountain lion will continue the attack until the victim has been killed. If a firearm is not available, pepper spray may be a viable alternative. Even considering its unknown effectiveness, it would seem to be preferable to no defense at all.

The decision whether or not to carry a firearm when working in areas where dangerous creatures are known to exist, ultimately rests with each member of the survey crew. The survey company may have rules prohibiting carrying firearms or it may have no rules at all. Some jurisdictions have firearm licensing requirements which must be complied with, particularly when individuals wish to carry handguns. Even if there are no prohibitions on carrying firearms, a member of a survey crew may have philosophical objections to firearms, fear firearms or have little or no experience with firearms. These considerations should be discussed before a crew embarks on a survey where crew members may be subjected to danger, so that every member will know what to expect, and be in a position to make a decision whether or not to participate in the work.

The decision whether or not to carry a firearm when working in areas where dangerous creatures are known to exist, ultimately rests with each member of the survey crew.

When considering whether to carry firearms, pepper spray or some other form of protection, it should be born in mind that when land surveyors work together in a crew, it becomes a joint enterprise. As anyone who has been a member of a group exposed to the threat of injury or death knows, each person has a moral duty, and in some situations a legal duty, to the welfare of his or her companions. A level of selflessness comes into play which transcends personal concerns and ideology. This duty must be the primary consideration when making decisions relating to the safety of the crew. Even when considering the low probability of an attack by a dangerous animal, if it does happen, you will not want to be in the position of having to explain to the family of an injured or deceased person, why you were unable to offer effective assistance. Surveyors are professionals who are held to a high standard of duty and care. Surely, part of this duty is ensuring that all members of the crew come home alive and well.

10.3 Safety When Working in the Field

The majority of dangers to which surveyors are exposed, occur in the field. Surveying along a highway is one of the most dangerous places a survey crew can work. One study showed that the risk of death in working alongside a highway is seven times higher for roadway workers than for other workers. Driver inattention is the single greatest cause of accidents. Other risk factors include bad weather, poor lighting and curves and hills which obstruct driver vision. The use of cell phones and texting while driving is increasingly dangerous to both pedestrians and other drivers. A study of cell phone use while driving claimed that holding a cell phone to the ear was not the real problem; the actual danger was caused by the lack of attention to driving caused by the driver being engrossed in conversation. Apparently, the human brain is not as good at multitasking as some of us might like to believe. Certain states have made the use of cell phones while driving illegal. In some states cell phones can be used, providing they are not handheld.

Texting is probably a greater problem than talking on a phone. In addition to the reduction in mental concentration on driving, texting requires a person to divert their vision from the road to the device. Texting by drivers creates pronounced danger to the driver and everyone else in the vicinity of the moving vehicle. There is no justifiable reason to do it.

Surveyors must be familiar with applicable safety laws and regulations which govern working along roads and highways in their jurisdiction. Precautions include wearing safety vests and deploying traffic cones and warning signs. Surveyors working along a road must remain vigilant and always have an escape route in mind, in the event an inattentive driver gets too close. Crew members should never turn their back to oncoming traffic. The person running the instrument is in particular danger because looking through the telescope in order to accurately sight the target, requires concentration. In addition, the person's eyes are not focused on oncoming traffic. If the instrument person must make measurements along a busy road, another crew member should be standing next to the instrument watching for traffic. Ideally, this person should wave an orange flag so that drivers are aware of the risk and steer clear. The person holding a prism pole locating evidence is similarly in a dangerous situation because he or she needs to remain focused holding the pole perfectly steady and plumb until informed that the shot has been recorded.

Surveyors are exposed to hazards even when working away from roads. When surveying in the woods, it is easy to slip when climbing over a wet log. Cutting survey lines through the woods requires using machetes, chainsaws and brush cutters. Hardhats, gloves, safety glasses and hearing protection should be used. If cutting trees is necessary, there are hazards from the tree falling or kicking out at the cut. Chainsaws are very dangerous tools because they are difficult to protect against. A chainsaw can seriously cut the user's leg in a spit second of inattention. Machetes can glance off a tree and go into a direction unanticipated by the user—possibly into the user's body. A strong blow with a machete into a rotten limb may have enough remaining energy to continue until it hits the surveyor.

Members of the survey crew must become well acquainted with plant species. Sometimes old deeds call for a certain type of tree as a boundary corner so a surveyor must be able to identify it. Being adept at plant identification can also eliminate a great deal of misery. A small number of plants growing in the woods are poisonous and survey crews must be able to identify them. The alternative is to become a casualty of the plant's poison. Poison Ivy, poison oak and poison sumac release urushiol, an oil, from the leaves when touched. The oil causes an allergic reaction of the skin called dermatitis, sometimes with severe itching. In bad cases the skin can blister. A serious case can incapacitate a person leaving them unable to work for days or weeks. Eastern poison ivy grows as a hairy, ropelike climbing vine on trees or brush, but it can also grow on the ground (Fig. 10.12). It is a very common plant and it can be found nearly everywhere in the woods, along fences, stone walls, along the edge of the road and even in yards. Western poison ivy is not a climbing vine. An old saying about poison ivy is "*leaves of three, leave it be*".

Fig. 10.12 Poison Ivy

Members of the survey crew must become well acquainted with plant species. Sometimes old deeds call for a certain type of tree as a boundary corner so a surveyor must be able to identify it.

Poison oak also has three leaves similar to poison ivy. Poison sumac is a woody shrub which may have yellow berries. Not everyone is allergic but most people can become allergic at any time even if not previously allergic—it is a matter of the amount of exposure. Wearing long sleeves, long pants, boots and gloves will help prevent contact. Indirect contact such as touching a machete or brush cutter that has been in contact with the plants is also likely to cause a reaction. So will touching clothing that has been exposed. Surveyors and others who walk in the woods should learn to identify these plants before inadvertently touching one of them and becoming infected.

In New England, when surveying along stone walls there is a temptation to run survey lines directly over the wall. There is more work involved in tree and brush cutting if survey lines are run alongside the walls. Walking on stone walls is very risky, particularly when the walls are made from rounded field stones which tend to be loose and roll underfoot. Balancing on the walls is difficult enough, but the real hazard comes from the wall collapsing. Many of the walls have been in place for 200 years on ground subject to continual freezing and thawing which can cause the individual stones to move into unstable positions. Trees or tree limbs often fall onto a wall and disturb the stability of the stones. It is not unusual to step on a stone and have the wall collapse underfoot. In addition to suffering the indignity of falling, a surveyor is likely to be injured by falling stones weighing a hundred pounds or more.

When surveying, it is often necessary to climb over walls. The previous cautions apply here as well. If it is necessary to climb over a wall, a person should never jump off the wall unless he or she is sure of the ground condition of the landing spot. Jumping into a pile of leaves covering a pile of stones is likely to result in a twisted ankle or broken leg, particularly if the jumper is laden with a backpack full of heavy equipment.

Another potentially dangerous situation arises when a steel tape is used to measure along an electric stock fence. Electric stock fences are installed to keep livestock enclosed. Electric fences can have voltages ranging from 3,000 to 10,000 volts. The voltage is not continuous but is only turned on for short intervals which are called pulses. If an animal or person touches the fence during a live pulse, the body acts as a conductor between the fence wire and the earth (ground) allowing an electrical current to flow through the body. The current causes a non-lethal but exceedingly painful shock. The equation $E = I * R$ (Voltage = Current * Resistance) gives the relationship between the voltage, current and resistance. For a given voltage, the lower the resistance is the greater the current will be. The lethality of the shock is dependent on the magnitude of current flow through the body. Body resistance can vary between 1,000 and 10,000 ohms depending on whether it is wet (sweating for example) or dry. So, the lower resistance of a wet body will increase the current flow and the degree of electrical shock. For humans, the electrical resistance of clothing, such as gloves and boots and whether they are wet or dry also affects the current flow.

As we saw in Chap. 4, it is customary when using a steel tape for the front person to carry the tape reel and let the fully extended tape drag on the ground while walking to the next point to be measured or set. In such cases, the rear person holding the free end of the tape must be careful to throw the tape clear of the fence when each measurement is completed, so there is no danger of the tape touching the electric fence. Like the steel wire in an electric fence, a steel tape has low electrical resistance so it is a very effective conductor of electricity. As the front person is walking with his back to the dragging tape, he is unable to keep an eye on the tape to be sure it does not touch the fence. It is therefore up to the rear person to ensure that the tape stays clear of the fence as it is dragged along the ground. If the tape does touch the fence, the person holding the tape reel is likely to receive a shock (and immediately discover that his or her boots are probably not very well insulated against such high voltages). Fortunately, the current of an electric stock fence is very small, and the pulse width very short, so the contact will be unlikely to cause any permanent physical harm. However the incident will certainly get the person's full attention.

The author has measured along miles of electric fence without a single accident. However, one incident does come to mind where two crew members were using a steel tape to measure a line next to an electric fence. The rear person - we won't divulge his name so let's just call him Charlie - was a skilled and experienced surveyor but he occasionally suffered from a puerile streak in his temperament. An introduction to these tendencies may help to acquaint the reader with Charlie before we proceed to describe the electric fence incident. One warm summer day, Charlie, while surveying a farmer's field stumbled upon a nest of unhatched chicken eggs, long abandoned by mother hen. Charlie could not resist picking up the eggs and aiming them at the party chief's legs. Fortunately, Charlie was such unskilled pitchers that his aim was well off the mark. The eggs did hit the ground close enough to the instrument that the smell of rotten eggs permeated the area for quite some time so that working there was rather unpleasant.

Returning to the electric fence incident, Charlie and another crew member were measuring alongside an electric stock fence. I was running the instrument and giving line to the front person by means of hand signals as the line was long and my voice would not travel such a long distance. Charlie's job was to hold the end of the tape as the distance was being measured and pick up the chaining pins as the measurements progressed along the fence. As soon as the first chaining pin was set and the front person started walking away, dragging the steel tape as he went, Charlie mischievously threw the end of the tape over the electric fence. For reasons unknown to Charlie, and to his surprise and dismay, the desired effect failed to materialize. The tape reel, being carried by its wooden handle, may have had a high enough resistance that it acted as an insulator and prevented the electrical current from reaching the intended victim. Or, it is possible that a portion of the tape may have fallen off the brush which was keeping the tape from touching the ground, and the short circuit prevented the current from traveling all the way to the person holding the reel. In any event, as the tape was being quickly pulled along, it immediately fell off the fence. Charlie, who could not be faulted for lack of resolve, persisted in throwing the tape over the fence, and after several attempts, Charlie's efforts finally came to fruition. Charlie's actions that day proved that perseverance is not always a virtue. Although, as I have said, the distance to the crew was substantial and they were unable to hear my instructions, I had no trouble hearing the exclamations and expletives of the afflicted crew member. Predictably, tempers flared. And for a while, Charlie's prospects of emerging from the incident unscathed did not look promising. The situation did cool down after a time and being in all other respects a likeable fellow, Charlie was fortunate in receiving only a verbal rebuke and a stern warning never to do it again.

Surveyors must always be prepared for rapid changes in weather, particularly if they are working far off the beaten path. In cold climates, and in mountainous areas, a rapidly moving snow storm could make it difficult for the crew to get back to the vehicle, especially if a lot of equipment must also be brought out. In hot climates, sunburn or sunstroke is always a possibility. Plenty of water and sunscreen should be carried. Surveyors should always check weather reports before heading into the field, so they can be prepared for changing weather. The survey crew should always carry a first aid kit as small injuries are relatively common. When working in the woods, insect repellant may be useful if the user is not allergic to it and can stand wearing it.

Clothing should be chosen appropriate to the conditions likely to be encountered. When working in the woods, the crew should wear boots, long pants and long sleeve shirts. Leather gloves should also be worn. The effects of mosquitos, poison ivy, briars and other, similar dangers can be either eliminated or greatly diminished by wearing protective clothing. When it is hot, there is an understandable reluctance to wearing long pants, long sleeve shirts and gloves. The uninitiated may even be convinced that they are exempt from being affected, or if they are affected, they can "handle it". Mosquitos, poison ivy and briars have a persuasive quality that derives from millennia of successful adaptation and they rarely fail to convince nonbelievers of their efficacy, usually in relatively short order.

The author is reminded of a situation in the mountainous wilderness of southeast Alaska in the summer. We were at an altitude of about 5,000 feet above a dense cloud cover with the sun shining and not a cloud above us. In the clear air it seemed like there were hundreds of mountains peaks poking through the white puffy cloud cover below us in every direction. As we were working and admiring the view, a hiker appeared wearing little more than shorts and a short sleeve shirt. Unsure of how to get back down the mountain, and not wanting to return the way he ascended, he asked if he could join our party in our descent to the main trail. As we followed the ridgeline, we descended into the dense cloud cover where visibility diminished to a hundred feet or less. As we moved down past the alpine, the trail became so steep that the only way to make it down the wet, slick rock was to grab onto the stems and branches of Devil's Club, which grew alongside the trail. Devil's Club is covered with spines on the stems and leaves. We were all dressed in jackets, long pants and leather gloves—except for our new acquaintance. In spite of efforts to help him, he suffered many cuts on his hands, arms and legs from the Devil's Club spines. By the time we got to the main trail, several hours later, his legs, arms and hands were covered in blood. Although painful, the lacerations were not life threatening and, in a few days, he was back to normal. The situation had the potential to become much more serious and he was fortunate that it did not. Circumstances like this can be easily avoided by dressing appropriately.

10.4 Land Surveyor Etiquette

Land surveyors are professionals, just like engineers, doctors and lawyers. A profession is a discipline that requires special education and training. Professionals are legally held to a high standard of ethics and conduct, and each state has regulations which govern the conduct of members of each of the professions. Unlike many other professions, surveyors tend to work outside, so they and their activities are highly visible to their clients and the public. As with any profession, it is the responsibility of the licensed professionals, and those working under their license, to represent the profession in a dignified manner.

When a client pays a boundary surveyor several thousand dollars to survey their property, they expect the surveyor to properly and carefully perform the work. In addition, they expect to be treated with courtesy and respect. They expect the survey crew to respect their property and, should the surveyor be required to enter upon the abutting property, to treat their neighbor and the neighbor's property with respect.

We have seen that surveyors generally do not perform field work alone and they are usually accompanied by one or more crew members. The survey crew has the same responsibilities to the client and the public as the licensed surveyor. After all, the crew members are working under the license of the professional land surveyor and they would not have a job if the surveyor were not licensed. If a member of a survey crew dresses inappropriately, acts in an unprofessional manner or damages the client's property, it reflects badly on the licensed surveyor and the firm.

Because surveyors are professionals, they should always dress appropriately. Clothing should be neat, clean and suitable for the type of work being performed. Sometimes, the field work may involve cutting brush or digging holes to set monuments so the crew members may get dirty or one of their garments might get torn—it may be unavoidable. However, this is not an excuse to begin a morning's work with dirty clothes or torn garments. Different types of work may require different types of clothing. Working in a swamp may require tall rubber boots or waders. It would not be desirable or appropriate to wear waders when performing the survey of a house lot in a residential neighborhood. Nor would it be appropriate for a crew member to work in such a neighborhood wearing beach party apparel. These examples are exaggerated; obviously, the purpose here is not to suggest what garments a survey crew should or should not wear, but to point out that the survey crew must be sensitive to the standards of the profession. The crew must show the level of respect that a client expects and deserves. A crew member that arrives at a residential work site dressed in sweat pants and worn out work boots with duct tape wrapped around the toes to hold the soles on, will place the entire survey crew in a bad light.

Some might argue that appearance should have nothing to do with it so long as the work is performed correctly. While in theory, this may be a valid argument, it displays a fundamental misunderstanding of the social context. Many clients will not have a clear perception of how the survey crew goes about establishing their boundaries. Unlike the days when traditional surveying techniques were used and a client was able to see the crew using a tape to make measurements along a boundary line, today we use beams of light and radio waves for positioning. Even in the old days, surveying using traverses and offsets, made surveying somewhat of an abstraction. The bottom line is that a client cannot see the measurements, so she will need to have faith that the crew is doing the best work possible. In order for the client to be comfortable with the crew's ability to make measurements and locate the boundaries correctly she must rely on other indications. Reputation of the surveyor is certainly one indication. A recommendation of the surveyor by an acquaintance is another indication. The appearance and actions of the crew are also important indications, because the client will be able to directly observe these for herself. If a crew member is disorganized, rude, disheveled or generally acts in an unprofessional manner, it will almost certainly leave a bad impression on the client.

If a crew member is disorganized, rude, disheveled or generally acts in an unprofessional manner, it will almost certainly leave a bad impression on the client.

Cigarettes should not be smoked while on a client's property (for health reasons, they should probably not be smoked anywhere). Soda cans, water bottles and candy wrappers should not be left on a client's property. Driving the survey vehicle across a client's manicured lawn because a crew member is too lazy to carry a concrete monument from the street to a back lot corner is sure to become an animated topic of conversation between the client and the survey crew. The use of music headsets by crew members is distracting and should not be allowed. When working near traffic, the music can mask the sound of approaching vehicles. It can also mask calls from another crew member. If for no other reason, it detracts from the concentration of the listener to the job at hand. It goes without saying that the din of jobsite radios is unprofessional and must not be used when the crew is on a client's property.

Cell phones can also be a problem. Crew members should generally not have cell phones on while working. No client paying an hourly rate wants to see a crew member talking or texting while on the clock. A possible exception is the crew chief, who may need to be in communication with the office. If a crew member has a family or other emergency, the survey office could be called and the message relayed to the crew chief.

Most surveyors use radios (Walkie-Talkies) for communication between crew members. This is clearly preferable to shouting across the client's yard. It must be kept in mind that the FCC has regulations and radio transmissions can be picked up by anyone having the right equipment. The party chief should insure that all radio users are familiar with FCC regulations:

> whoever utters an obscene, indecent or profane language by means of radio communications shall be fined not more than $10,000 and/or imprisoned not more than two years. (Title 18, US Code, Chap. 71, paragraph 1464)

Surveyors have traditionally made use of hand signals for communication between crew members. Hand signals can be used, for example, to tell the rod person to go left or right or "OK". These signals are more or less universal and they should be learned by all new crew members. Sometimes hand signals are more economical of time than having to hold a data collector in one hand and a radio in another. Some surveyor's safety vests and jackets have a provision to locate a radio or microphone high up near the head so it is not necessary to hold the radio. Some radios are voice activated. Other radios have a "call" function which transmits a short beep which can be used to signal an "OK" when the instrument person has taken a shot. This can save the instrument person's voice at the end of a long day and it is a relief from the monotonous transmission of "OK" when taking hundreds of consecutive shots.

Total stations are expensive. A robotic instrument might cost $30,000 or more. Some surveyors have a rule that the transit person is never to leave the instrument unattended, particularly if it is in a populated area where it could be stolen or tipped over. Often, crew members are out looking for evidence or trimming brush which leaves the transit person with nothing to do but wait at the instrument. Under such circumstances it is prudent that the transit person not sit or lay on the ground while waiting. No client wants to arrive home and find a crew member lying on her lawn while she is paying the crew a couple of hundred dollars per hour. This advice is applicable to all crew members.

The survey vehicle should be kept reasonably clean and tidy. Clients are not favorably impressed when a crew member opens the vehicle door to exit and a couple of empty soda cans (or worse) roll out onto the sidewalk. Surveying is a precise profession. An untidy truck is a sign of sloppy habits. Many surveyors assign one of the crew members, sometimes the newest person on the crew, as the equipment- person. This person is responsible for ensuring that, at the end of the day, none of the equipment has been left on a job site, that the equipment is in good repair and that damaged or lost equipment is noted for replacement. The person might also be responsible for keeping the truck clean and serviced. It is embarrassing to return to a job site to find that a tripod and prism are still set up over a traverse point—left there from the previous day. It is more embarrassing to have to explain to the company owner that the tripod and prism were left at the site, but the equipment is no longer there.

It is embarrassing to return to a job site to find that a tripod and prism are still set up over a traverse point—left there from the previous day.

It has already been noted that total stations, particularly servo driven and robotic total stations are very expensive and can cost upwards of $30,000. These instruments are also very delicate and they should be handled with great care. We learned that the instrument should be removed from the tripod when setting up and only placed in the tribrach after the setup is complete. We live in a world so replete with technology that it is easy to take for granted the cost and value of equipment. Nevertheless, surveying equipment is expensive and it should always be handled with care. The fact that survey crews often work under difficult environmental conditions makes it even more imperative that equipment be handled carefully. When setting up a tripod, there is no need to kick one of the legs away when it could just as easily be moved by less forceful means. There is no need to throw the instrument case on the ground, when it could be just as easily placed on the ground. Prisms poles and prisms should be kept in accessible containers in the survey vehicle, not thrown in the back of the truck along with the shovel and iron bar. If the survey crew exhibits care in the handling of equipment, this care is likely to transfer to the level of care necessary to make precise measurements. Careless habits are likely to result in careless work.

Chapter 11
Field Work—Finding Monuments

The first order of business when the surveyor goes into the field on a new survey is to find the monuments that will control the locations of boundaries. These monuments could be street monuments which fix the location of street lines. They could be monuments called for in the client's deed. They could be monuments called for in the deeds of abutting properties. If the client's land is a lot in a subdivision, the monuments could be called for on the subdivision plan.

11.1 Monuments Must Be Called-for in the Record in Order to Control

When a surveyor finds a stone bound or other monument near a property corner there is an understandable tendency to consider that monument as controlling the location of the property corner. However, for the monument to be controlling it must be Called-For in the record information. The monument must be an original monument. The deed describing the parcel, or a plan called for in the deed, must specifically call for and describe the monument. The monument as found must comport with the record description as to location and type. For example a deed call such as "thence running North 34° West, 456.34 feet to a stone bound" would be a call for the stone bound. If the deed call was instead "thence running North 34° West, 456.34 feet to a corner" and a stone bound was found at that location, the bound would not control the location of the record lot corner. An exception to this rule is that a bound not called for in the record would be controlling if there were other evidence to show that it actually occupied the record location.

P. Gay, *Practical Boundary Surveying*, DOI 10.1007/978-3-319-07158-9_11

For the monument to be controlling it must be Called-For in the record information.

It is important to realize that a call in a deed for a monument is not actually a call for the monument itself; it is a call for the *location* of the monument. This makes sense if one considers how easy it is for original monuments to become disturbed. For example, a stone bound set in the woods could be tipped over or upended by a falling tree. A drill hole in a stone wall could be moved by a person rebuilding the wall. Monuments like iron pipes or rebars are easily pulled out of the ground and moved. Merely finding an original monument is not enough. The surveyor must determine that the monument has not been disturbed. In actual practice, many old monuments show signs of having been disturbed.

The surveyor must be experienced and knowledgeable in monument identification and scrutinize a found monument in order to determine if it is indeed the original monument. This is a situation where there is no substitute for experience on the part of the surveyor. It is an example of a skill that cannot be perfected without considerable exposure to field work. Experience and familiarity with the practices and customs of local surveyors, perhaps long deceased, is essential.

Some examples may help to illustrate the point. A concrete bound is not a stone bound. Sometimes the two appear deceptively similar, particularly when the bound is buried under 3 feet of fill and the surveyor is peering into a dark hole with a flashlight with water seeping in faster than the surveyor can bail it out. A copper pin in a stone is not the same as a brass pin in a stone, even though oxidation over many years makes it look much the same and it appears to be at the correct location (Fig. 11.1). Three hack marks in a white oak tree are not the same as three hack marks in a black oak tree. Surveyors must be experts at plant identification. Sometimes the cavity left by a small stone falling out of a conglomerate stone looks very much like a drill hole. In this case, you would probably ask yourself why anyone would put a drill hole in a conglomerate stone in the first place, knowing that it would be almost impossible to recover it with certainty at a later time. If a deed written in 1815 calls for the corner to be a wooden fence post what is the probability that a fence post found in 2012 is the same post? Unless it happens to be in an extremely arid climate, in all likelihood the original post rotted out long ago.

One must never discount the possibility that the record information has described a monument incorrectly. This occurs more often than one might imagine. In such cases, a boundary surveyor needs to think like a sleuth. Being familiar with Sherlock Holmes mysteries sometimes helps to instill the necessary level of inquisitiveness. As an example, in a survey performed by the author, the current typed deed called for the property line to begin a certain fixed distance northerly from a "wall". The property owner inherited the property from his parents and the family had lived on the land for many years. When asked about the location of the wall, the property owner stated that, to his knowledge, no wall had ever existed at that location. An initial site investigation disclosed no visible evidence that a wall had ever existed there.

Fig. 11.1 An old brass pin found in a stone wall

> One must never discount the possibility that the record information has described a monument incorrectly.

The only way of fixing the boundary for this parcel of land was to somehow discover where the original wall was located. Another trip to the registry of deeds was made, and the deed was traced back to a handwritten deed created in the early 1800s. For those unfamiliar with ancient handwritten deeds, some of the professional penmen in early U.S. history used beautiful flowing scripts with particular flourishes. The writing was ornate and sinuous. Although most children are not taught the art of penmanship these days, in an earlier time it was considered to be very important to education and writers took their writing very seriously. Today, readers who are not proficient in reading these ancient writings must be careful when interpreting the words in the original script.

Getting back to our example, when the early handwritten deed was carefully reexamined, it was discovered that the original word was "well" not "wall". Upon close examination with a magnifying glass (as Mr. Holmes would have done), the letter was clearly an "e" not an "a". The person who drafted the conveyance subsequent to the original one misread the original deed and wrote "wall" instead of "well". The error continued to be perpetuated in later conveyances right up to the recent, typed deeds. This was clearly a typographical error because the language before and after the error was identical in all respects to the more recent deeds. Nothing else in the deed description had changed in the almost 200 years since the original deed was written. The error was perhaps understandable, because walls were, and still are, much more commonly used in describing boundaries than are wells so the person who drafted the incorrect deed probably expected to see the word "wall" not "well".

In fact, there was a well located on the property exactly where the "wall" should have been. It was an old hand dug well lined with field stones. There was no reason to suppose that the existing well was not the original well called for in the ancient deed. Mystery solved. This example helps to illustrate how careful surveyors must be in interpreting and researching record documents. A single letter made the difference between being able to tell a client where his boundary was located and not being able to locate the boundary.

11.2 Kinds of Physical Monuments

Monuments were discussed in general in an earlier chapter. Here we discuss a few of the types of physical monuments commonly found. The identification of physical monuments is sometimes quite difficult and some monuments would pass unobserved by nearly everyone except someone experienced in their identification.

Stone walls are ubiquitous in the northeast U.S. The author has been asked on many occasions why there are so many stone walls running through the woods. Why would anyone want to build a stone wall in the woods? As New England became settled during the 1700s, people supported themselves through agriculture. Farmers needed to keep their animals from entering a neighbor's land and damaging crops. If this were to happen, a farmer would be liable to their neighbor for damages. Before stone walls were built, wooden fences were used to contain livestock and horses. As more land was cleared, wood became less available. Wood fences also rot over time so they required constant maintenance. Because of the overabundance of stones in fields and woodlands, stone walls became the obvious solution. They solved two problems. First, they were a way to dispose of some of the stones littering the fields which made cultivating for crops time consuming, difficult and hard on equipment. Second, the walls provided a permanent and low maintenance replacement for wood fences. Stone walls eventually replaced most of the wood fencing formerly used along the perimeter of fields and open meadows which were used for crops and for livestock grazing.

Many stone walls were built in New England during the late 1700s and early 1800s. In addition to enclosing livestock, some of the walls served as boundary markers. If a farmer's field bordered a neighbor's field, it was sensible to use the wall as both a means of enclosure and as a boundary marker delineating the limits of ownership of each of the properties.

After 1935, as more and more people began to leave farms to work in cities, the number of farmers making a living on small farms began to decline. As a result, the fields became overgrown and eventually turned into woodlands. This is why there are now many stone walls running through the woods. Some of these walls are physical evidence of boundaries. The reader should keep in mind that if a deed was written 200 years ago there is always a possibility that the original wall could have been moved or obliterated.

Where stone walls are used to establish boundaries between abutters, the boundary line runs down the middle of the wall. Each property owner therefore owns half of the wall. In some jurisdictions removal of a boundary wall, where ownership of the wall is shared between the abutters, is a civil and/or criminal offense. In some cases, a boundary running along a stone wall may not be the middle of the wall. The rules vary by jurisdiction. Where walls run along highways, and there is no formal highway layout with takings, the line may be on the street face of the wall. This places the wall entirely on private property. The same rule may apply to a cemetery where the line would be on the outside face of the wall so that the wall was entirely within the cemetery.

> Where stone walls are used to establish boundaries between abutters, the boundary line runs down the middle of the wall.

Drill holes have been commonly used as property corners and the record is replete with calls for them. Sometimes drill holes are in stone walls (Fig. 11.2). In other instances they are in stones or in stone or concrete bounds. In the latter two cases once the stone or concrete bound has been found, the drill hole should be obvious. Drill holes in large stones and in walls can be very difficult to find, particularly if the drill holes are small. Small drill holes tend to get filled up by mud placed by insects. Dirt and lichens also obscure drill holes. It may be thought that small drill holes are a little more accurate or can be more accurately placed than larger drill holes, but this would seem to be small advantage if the drill hole cannot be found by the property owner or surveyor. There are stories of an old, ornery surveyor who had a habit of covering up his drill holes with other stones to make it harder for others to find them, thereby making him the only one who knew where

Fig. 11.2 An original drill hole set in a stone wall

they were located. Whether or not this is true and whether or not this would result in more repeat business for the old surveyor is unknown. The fact is that many drill holes are difficult to find, even when one knows exactly where they ought to be. Figure 11.3 shows a drill hole in the side of a stone found buried beneath the leaves covering the forest floor. As there are sometimes hundreds of stones in the vicinity of a property corner, often the only way to find a drill hole is to calculate its probable location and use an instrument to turn an angle and measure a distance to where it should be.

Copper or **Aluminum Pins** are much easier to find than drill holes and for this reason alone, surveyors should consider using them instead of drill holes

Fig. 11.3 Drill hole found buried under leaves

(Fig. 11.4). Pins take a few minutes longer to set than a drill hole but the result is well worth the small extra time and effort. In order to set a pin, a hole is drilled in a stone and a malleable and durable metal such as copper or aluminum is hammered into the hole so that the metal expands and becomes wedged in the hole. Because the pin sticks up above the surface of the rock it is generally much easier to find than a drill hole. Many of the older surveyors used copper or iron pins to mark property corners but because of the time and economic pressures imposed by the modern world, this practice has unfortunately diminished in recent years. It should be recognized however that, in retracement surveys, the time expended in attempting to find original drill holes results in costs which far exceed what it

Fig. 11.4 A copper pin set in a stone wall

would have cost to set a pin during the original survey—sometimes by an order of magnitude.

Stone bounds have been used for centuries for property corners and they are found in all shapes and sizes. Sometimes stone bounds were made by splitting and chipping stone, such as granite, into a four sided post several feet in length. Other times a natural stone having a more or less square cross section was found and used as a property corner. Occasionally, a natural stone having a triangular cross section, similar to a tetrahedron, was used. In many cases, where natural stones were used, it can be difficult to tell the bound from other naturally occurring stones in the area. Figure 11.5 shows a typical granite stone bound.

Stone bounds can be two or three feet in length to six or eight feet in length. It is not uncommon for the top of old stone bounds to be 4 or 5 feet out of the ground. Such bounds often tip, over the years, to one side or the other. Except for state or municipal highways, concrete bounds are now used almost exclusively, particularly in the private sector. They are generally four to six inches square. So as not to be moved by the ground freezing and thawing, the bound should be long enough so that

Fig. 11.5 An old stone bound

the bottom is well below the frost level. Today, bounds ranging from three to six feet are used. If a bound is to be set near a driveway, sidewalk or in a lawn it is often buried so that the top is flush with the ground or a slightly above ground.

An advantage of concrete bounds over stone bounds is that they contain **Steel Reinforcing Bars** (rebars) so that if buried, they can be found with a metal detector (Fig. 11.6). A disadvantage of concrete bounds with rebars is that if they are hit by a machine such as a snow plow or dozer blade, the rebar tends to hold the bound together causing it to move substantially. Sometimes the entire bound may be yanked out of the ground. Stone bounds, or concrete bounds without rebars, may snap off at ground level leaving the base in its original location. This is more likely if the ground is frozen.

Fig. 11.6 Original round concrete bound found buried near the edge of pavement. Steel rebars allow bounds to be found using a metal detector

Stone or concrete bounds often have a drill hole or pin in the top. Sometimes a hole is drilled then filled with lead and a tiny brass escutcheon pin is driven into the lead. Modern surveying techniques can locate boundary corners to about 1/8 of an inch, so it is now common to mark the top of the bound in some way in order to maintain accuracy, otherwise it would not be possible to know exactly where the property corner is located on the top of the bound. In days past, when surveying was less precise, stone bounds were often not marked. If you run across an unmarked bound the convention is to use the center of the top of the bound as the location of the corner. An exception to using the center of a non-marked bound is sometimes used with highway bounds. In this case, the center of the back face of the bound is used, thereby placing the bound entirely within the road layout. This practice may not be universal and may vary in different localities according to local custom.

If a bound is obviously leaning because of some past disturbance, the surveyor should measure the amount of displacement and adjust the property corner accordingly. Recall that the property corner is fixed by the original location of the monument, not the monument itself. In our example, the bound is merely evidence of the original location. If the bound has obviously moved from its original location, then it cannot be at the property corner. The field notes should contain a record of the measurements made to the disturbed bound. In such cases it is prudent to photograph the bound and keep the image with the survey files. The surveyor should always note the distance and direction of the displacement on the plan or survey report. In cases where a bound has been disturbed, a surveyor is actually exercising judgment about where the original monument was located. An example of such a notation might be: DH in SB (found), N44°59′54″W, 0.36′. This tells the reader that the drill hole in the bound was found 0.36 feet northwest of the property corner. In other words, the location of the actual property corner has been adjusted to be at the presumed location of the original monument. Another surveyor may not agree with the interpretation and may choose a different location for the corner.

If a bound is obviously leaning because of some past disturbance, the surveyor should measure the amount of displacement and adjust the property corner accordingly.

Sometimes, a surveyor may have ties to the original location of a monument. Ties are simply distances, usually measured with a tape, from the monument to other permanent objects. For example, ties are often used when laying out streets and highway monuments and baselines. If a house with a concrete foundation were located close to the monument, measurements could be made from two corners of the foundation to the monument. Because a house foundation is relatively permanent, if the bound were to become disturbed, its location could be re-established by measuring from the foundation.

Some deeds call for a corner to be marked by a **cedar** or **locust post** or a post made from some other long-lasting wood. Because these woods are quite durable, depending on the climate and soil conditions, many are in existence 100 years after they were set. Wood posts were sometimes set in a **pile of stones**, so that if the post were to rot away, the pile of stones would continue to mark the location.

Piles of stones were also used without wooden posts (Fig. 11.7). Sometimes, long boundary lines were marked at intervals with them. Often they are not piles at all, but simply a series of stones placed on the ground in the form of a ring, two or three feet in diameter. Over the years, the stones tend to become covered by plant growth and debris. Natural processes eventually bury or partially bury the stones making them difficult to find. In areas where there are many naturally occurring stones, it can be challenging to tell the difference between a man-made pile of stones and natural stones. In such cases it helps to have intelligent feet. Because the stones often rest on other stones beneath them, sometimes standing and rocking on the stones allow you to feel and hear the loose top stones rolling on stones below them. The stones make a sort of rumbling sound. If you try this, be forewarned that bystanders—or worse, your client—may find it a bit strange to see a surveyor dancing in the woods.

Many old deeds call for trees as property corners. If the deed is very old the tree may be long gone. Sometimes the stump of a large tree remains. Deeds, on occasion, will call for a wire fence as a boundary line. Many old wire fences were made by nailing wire to trees in order to economize on fence posts. It is not

Fig. 11.7 An old stone bound set in a pile of stones. The iron pipe is a witness marker used to make it easier to find the boundary corner

unusual for the posts to be gone, while pieces of wire may still be found protruding from trees as in Fig. 11.8. When boundary fences were made by nailing wire to trees, the property lines will probably meander quite a bit.

In times past, trees were often marked as a method of indicating property boundaries or corners. Figure 11.9 shows such a mark. These marks are referred

Fig. 11.8 An oak tree that has grown around an old wire fence. Notice the many stones on the ground

to as a **Tree Blazes**. Sometimes the wood below the bark is carved with an inscription describing the line, corner, property owner or other information.

11.3 Finding Monuments—Initial Steps

Depending on the circumstance of a particular survey, the surveyor may prefer to go into the field and look for monuments before actually beginning precise measurements. One advantage of this approach is that the surveyor will then know exactly where to run the traverse lines. Finding monuments can often be done

Fig. 11.9 Tree Blaze in a forest in Alaska

using a tape and a compass or handheld GPS receiver. In other cases this may not be feasible and it is necessary to actually run survey lines to points somewhere near where the surveyor believes the monuments should be located. The probable position of the monuments can then be calculated so that the search area is narrowed and the crew can return to the field and look in the designated area.

Sometimes, particularly in wooded areas, it is necessary to find monuments before cutting line so that the survey crew will be sure they are not cutting on the abutter's land. In other cases, where for example, the boundary is a stone wall, the surveyor has a pretty good idea of where the property lines are and the lines can be safely cut. In many circumstances a few original monuments are initially found and, from these monuments, the surveyor is able to calculate the precise location of the remaining monuments. A second trip to the field should then determine whether or not the remaining monuments are still in existence.

11.4 Diligence in Finding Monuments

Boundary surveyors have an ethical obligation to be diligent in finding original monuments. During a retracement survey, the surveyor must find the original monuments used or set by the original surveyor. When performing the survey of a parcel on a public or private way, the surveyor must find monuments which establish the location of the way.

> Boundary surveyors have an ethical obligation to be diligent in finding original monuments.

Finding monuments is oftentimes difficult. If the original survey is an old one, some monuments will have probably been destroyed. Others may be found disturbed, making their original locations uncertain. Monuments are destroyed or disturbed through natural causes and by human activity. Wooden stakes rot over time. Iron pipes and rebars rust through. Stone and concrete bounds get run over and broken off by snow plows and heavy equipment. Drill holes in walls become disturbed as the stones in walls move by the action of frost and by trees and limbs falling on the walls. People sometimes take stones from a wall to use in a construction project, not knowing a stone contained a boundary marker. Some monuments get covered over by fill or by deposition from flooding. Monuments along shorelines become disturbed or lost as a result of erosion and moving ice sheets. Earthquakes, such as transform faults, cause locations on the earth's surface to shift taking boundary markers along with it.

In the future, we can probably expect boundary markers to become more subject to destruction than in the past. As climate change causes more severe weather, sea levels will rise causing encroachment along water boundaries. Hurricanes and tornados will become more severe and more frequent. The high winds and water associated with these events will take their toll on boundary markers. So will the clean-up efforts by municipalities and government agencies. When equipment operators are working under stress in order to save lives, prevent injuries and protect property, being careful to avoid hitting a not-too-visible boundary marker is probably not high on their list of priorities.

One way to help preserve boundary locations is to use GPS to tie boundary corners into state plane coordinate systems. It should not be supposed, however, that the use of state plane coordinate systems eliminates the need for setting permanent markers on site. It merely makes re-establishment of lost markers easier than it would be without it.

In spite of the difficulties in finding original monuments, a surveyor has a duty to ensure that the monument being sought is found or, alternatively, to make a determination that the monument is actually missing or disturbed. Merely walking to the presumed location of a monument and "looking around" is not good enough. The process of finding monuments can be very time consuming and labor intensive and this translates into expense. Surveyors, when estimating the cost of a survey, need to take this into account and inform their client of the importance of finding monuments, even if the survey ends up costing more.

In spite of the difficulties in finding original monuments, a surveyor has a duty to ensure that the monument being sought is found or, alternatively, to make a determination that the monument is actually missing or disturbed.

Ethically, there is no excuse for not exhausting every avenue in finding original monuments. Even when there is evidence, or reason to believe that a monument may have been destroyed, it is the surveyor's responsibility to make a determination as to whether or not the monument is still in existence. Many people would be surprised to learn how many old markers can be recovered if you know how and where to look and if you are persistent. In order to be successful at finding old monuments, a surveyor only needs a few essential tools. Chief among these are determination, perseverance and a good shovel. A few examples from the author's experience may help to illustrate the point.

In one case, a deed called for a stone wall. However, the area where the wall was supposed to be located was a large lawn. The property owner of 30 years could not remember ever seeing a wall. A probe was used in the area where the wall should have been and it became apparent that there appeared to be a line of stones just below the surface. A small backhoe was hired and the area carefully excavated. Figure 11.10 shows the wall base clearly visible below the surface of the ground.

Figure 11.11 shows the base of a stone bound found after a coastal hurricane. The bound was located on a beach, a hundred feet or so from the ocean. It was likely broken off through the negligence of a heavy equipment operator clearing debris from the road running along the beach after the hurricane. The base was found almost two feet below the surface. Looking closely at the monument a coloration showing the depth of burial is just visible. To an inexperienced observer viewing the monument lying on the ground, it might seem that the bound was simply tipped over and not broken off. However, to an experienced observer, the length of the buried area, as measured from the discoloration, is far too short to have supported such a tall monument. The length above the discoloration was about 5 feet. The base of the monument would have had to extend at least four feet below the surface in order for the bound to have survived the 200 years that it did, particularly in such a severe environment at the edge of the ocean where it was subject to flooding and debris from hurricanes. Although not apparent in the image, the broken face also showed signs of being newly broken. Excavation revealed the base at its original location.

Fig. 11.10 Stone wall found buried just below the surface

Remember, it is the **original location** of the monument, not the monument itself that controls the location of the boundary corner.

Sometimes it seems very unlikely that an original monument is still in existence. Where a relatively new sidewalk was located in exactly the spot where a stone bound should have been, it seemed likely that the bound had been destroyed when the sidewalk was installed. As shown in Fig. 11.12 the stone bound was found located beneath a sidewalk in an urban location. It was found by using a long probe angled in from the grassy area just outside the sidewalk. The probe contacted a hard object. Once the area around and below the sidewalk was excavated, the monument was found. Normally a monument such as this would have

Fig. 11.11 Stone bound base found buried, its location undisturbed

a drill hole centered on the top. Here the sidewalk was poured directly over the bound which obscured the drill hole. So, rather than destroying the sidewalk (and possibly being prosecuted by the municipality, for the destruction of public property) the soil was excavated as much as possible around the monument, thereby determining its size. It was then a simple matter to calculate the approximate location of the drill hole from the faces of the monument. Of course, this reasoning relied on the presumption that the original drill hole was centered on the top of the bound—not always the case. However, under the circumstances, the best that could be done was to measure to the sides of the bound and note the plan that only a "bound" and not a "drill hole in a bound" had been found, along with a note that the bound was found under the sidewalk.

Fig. 11.12 Stone bound found under a sidewalk

As the images in this chapter show, monuments are often found buried below the ground surface. There are many reasons for this. Land is sometimes re-graded after a monument is set. This often happens with new construction where the monument is set before a lawn is graded and planted with grass seed. If the monument is an iron pipe or rebar, property owners may be concerned about hitting the metal object with a lawn mower or having to constantly mow around it so they hammer it into the ground. Iron pipes and rebars protruding from the surface are also a hazard to children and pets running around a yard. A stone bound may get hit by a vehicle and break off at the ground surface. Sometimes when a surveyor begins to dig a hole to set a concrete bound, a large stone or ledge that is too large to move is discovered just below the surface, so the only solution is to install a pin in the ledge and fill in the hole. Buried markers can be a challenge for surveyors to find (Fig. 11.13). If the marker is ferrous, a metal detector can be used. Metal

Fig. 11.13 An original iron pipe found buried

detectors will find any steel or iron object so, in places such as industrial sites where they may be lots of metal pieces buried in the soil, a metal detector may not be particularly useful. Buried metal, fence pieces, sewer pipes, water pipes and other metal objects will cause metal detectors to give false readings. Figure 11.14 shows a buried metal monument. Finding buried monuments in areas where there are manicured lawns is particularly problematic as a surveyor cannot just dig up the lawn unless he or she is relatively certain that the sought after monument has been found.

When monuments are located below the surface of the ground a long steel probe can often be pressed into the ground to find buried evidence. A surveyor experienced in using a probe can often gather quite a bit of information about an object below the surface with judicious use of the probe. For example, the size and shape of the object can be determined simply by continued probing of the area. Most concrete and stone monuments are only 5 or 6 inches square so the penetration of a probe will instantly go from a fixed depth while on the bound surface to a much lower resistance area where the probe passes by the bound's edge. The depth can also be easily determined by noting how far down the probe penetrates before encountering resistance.

Although the material from which the object is made cannot readily be determined with a probe, its mass can sometimes be inferred by the sound and feel of the probe. Hitting a small stone with a probe sounds and feels quite different than

Fig. 11.14 Iron pipe found buried below the ground surface

hitting 4 foot stone monument rigidly anchored in the soil. In fact, forcing the probe into the ground too aggressively can cause the user to experience a noticeable, and sometimes painful, shock when it fetches up against the stone. Buried stones often have very irregular surfaces and it is possible to determine this with the probe. Sometimes, as in the previous sidewalk example, a probe can be used to penetrate the ground at an angle, thereby helping to discover objects underneath impervious surfaces. One great advantage of the probe is that it only makes tiny holes in the ground so when it becomes necessary to look for a monument under someone's flawlessly kept lawn no damage is done.

If the probing results are promising, a shovel can be used to dig down to the object. A small whisk broom of the type commonly used with dust pans is handy for cleaning of the surface of the buried object, for example, to see if it contains a drill hole. When looking for monuments in areas where people have nice lawns, the survey vehicle should have a small canvass tarp and a narrow shovel such as a drain spade or trenching shovel. This allows a very small hole to be dug. First, the sod should be carefully removed in one piece and set aside on the tarp. The remaining soil can then be placed on another area of the tarp. If done carefully, the monument can be located, the hole refilled and the sod replaced in such a way that it would difficult to notice any disturbance of the lawn. While the hole is still open it is a good idea to paint the monument with fluorescent paint and run a piece of surveyors orange ribbon from the monument to the ground surface while backfilling. If another surveyor were to look for the same bound, the tip of the probe may bring back some of the orange paint. As soon as anyone starts digging they will see the orange ribbon and know they are on the right track.

Surveying in winter, when there is snow cover, makes the process of finding monuments considerably more difficult. In some cases it would be best to postpone the survey until after the snow has melted, although this does little to help a surveyor pay the mortgage during long winter months. Figure 11.15 shows a bound buried beneath the snow. Its location would have been obvious without snow cover. Finding monuments under these conditions can be difficult unless the surveyor knows exactly where to look. Frozen ground presents an additional difficulty because it makes it more difficult, if not impossible to use a probe and shovel to dig for evidence.

Fig. 11.15 Bound buried beneath the snow

Fig. 11.16 A trace monument—a monument which is no longer there but evidence of its location remains

Occasionally, the sought after monument is missing but there is evidence of its former location. These monuments are referred to as **Trace Monuments**. Figure 11.16 shows an area of concrete which was poured around a wooden post which originally marked a property corner. Over the years, the post rotted out completely but the concrete marks its original location. Notice that the concrete is well below the surface of the ground so it would not have been visible without digging. Here again, perseverance and a good shovel, helped to achieve a successful outcome.

Chapter 12
Field Work—Traversing

Boundary surveying often makes use of a traverse, particularly for larger parcels. One might imagine that the easiest way to survey a parcel of land would be to set the instrument on one property corner, point the telescope to a previous corner, turn the record angle and measure the record distance to set the next corner then repeat the procedure for each successive property corner.

This can be visualized by looking at Fig. 4.6. Suppose that lot corners "A" and "D" were in existence and the client wished a surveyor to set the remaining corners "B" and "C" which were lost because of some excavating in the area. A surveyor would set the instrument at property corner "A", point the telescope at corner "D" then turn the angle $98°40'35''$ and measure the distance 141.42 feet to set corner B. The surveyor would then set the instrument at corner "B", point the telescope at corner "A" then turn the angle $104°08'14''$ and measure 121.63 feet to set corner C. As a check, a thorough surveyor would then set the instrument on corner C and repeat the procedure to see how close he came to corner D. The difference between the point so set and the original corner "D", the error of closure, would tell the surveyor if the lot closure was acceptable.

In reality, the surveying technique just described is rarely possible or even desirable. In many cases, property lines are where people plant shrubs, trees, build walls and fences and sometimes pile junk to get it out of the way. These property line enhancements make it unlikely that a surveyor would be able to see or measure along the property lines. Even in cases where it would be possible, features such as crumbling stone walls make it dangerous to attempt to walk on the wall.

> In many cases, property lines are where people plant shrubs, trees, build walls and fences and sometimes pile junk to get it out of the way. These property line enhancements make it unlikely that a surveyor would be able to see or measure along the property lines.

P. Gay, *Practical Boundary Surveying*, DOI 10.1007/978-3-319-07158-9_12

In our simple scenario, boundary markers were conveniently found at corners "D" and A". Oftentimes the only remaining boundary markers are not on a common boundary line but are well separated and it is not possible to see or measure between them. For example suppose the only remaining markers in Fig. 4.6 were at "A" and "C" and the client wanted corners "B" and "D" set. If there were a building blocking the view between "A" and "C" it would not be possible to directly use these points to set the remaining corners.

Surveyors get around such problems by "running a traverse". A traverse is simply a series of survey lines that are run at convenient locations, usually not too far from actual boundary lines. Consider Fig. 12.1 which shows the same parcel of land, as in Fig. 4.6, but this time with a traverse. The traverse is the red dash-dot line. In this scenario the surveyor was required to set lot corners "B" and "D". Corners "A" and "C" were in existence from the original survey in the form of drill holes found in concrete bounds [DH CB (fd)].

The actual location of the traverse lines is chosen by the surveyor after considering a number of factors. For example, because the instrument must be set on each of the numbered points (1–5) it must be possible to easily and quickly see and measure

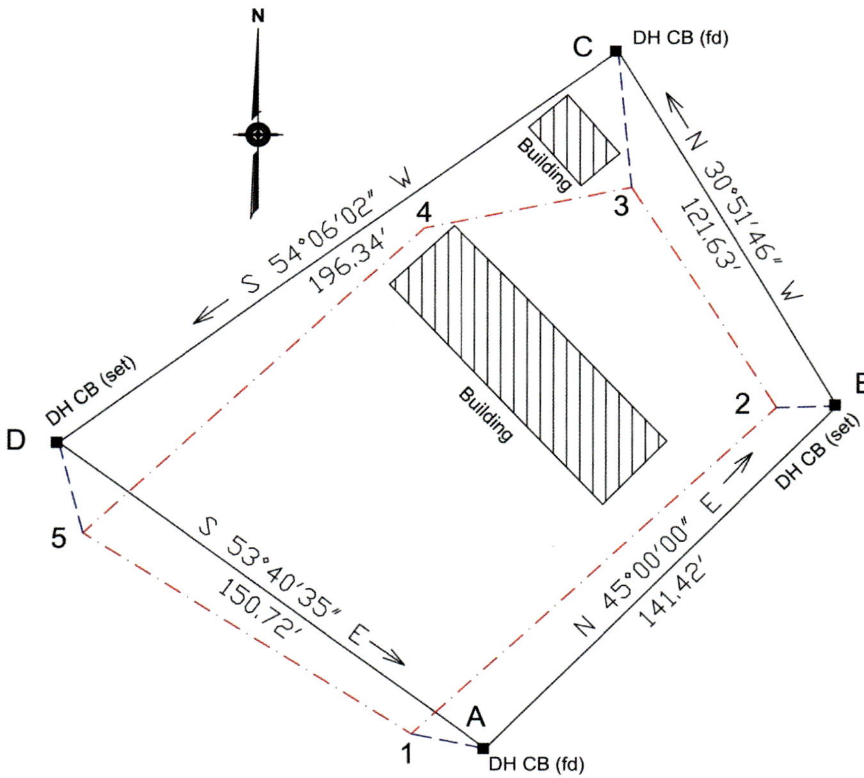

Fig. 12.1 A closed traverse

between these points. A technical way of saying this is to say that the points must be **Intervisible**. Intervisible simply means that you can see between the two points. If there were a tall rose bush between points 1 and 2, it would be necessary to move one or both points in order to see by the bush, unless the property owner wanted the bush cut down. Another consideration is that the ground where each point is located should be firm and relatively level so that it is easy and convenient to set the instrument there. If one of the lines were to run along a street, the surveyor would want to take safety into consideration and maybe run a line down a sidewalk instead of in the traveled area of the street.

A third consideration is that a surveyor tries to set the traverse points somewhere near points which are to be located or which are to be set. In this case, points 1, 2, 3 and 5 are all relatively near the lot corners and each of the lot corners is visible from these points. The blue dashed lines represent the survey lines from the traverse points to the points to be located or set. Traverse point 4 was needed because a building prevented seeing and measuring between point 3 and point 5.

A traverse is subject to the same rules of closure as the property lines. In the example, it would have been possible just to use points 1, 2, 3 and 5 and not use point 4. This is illustrated in Fig. 12.2. Each of these points is near a lot corner

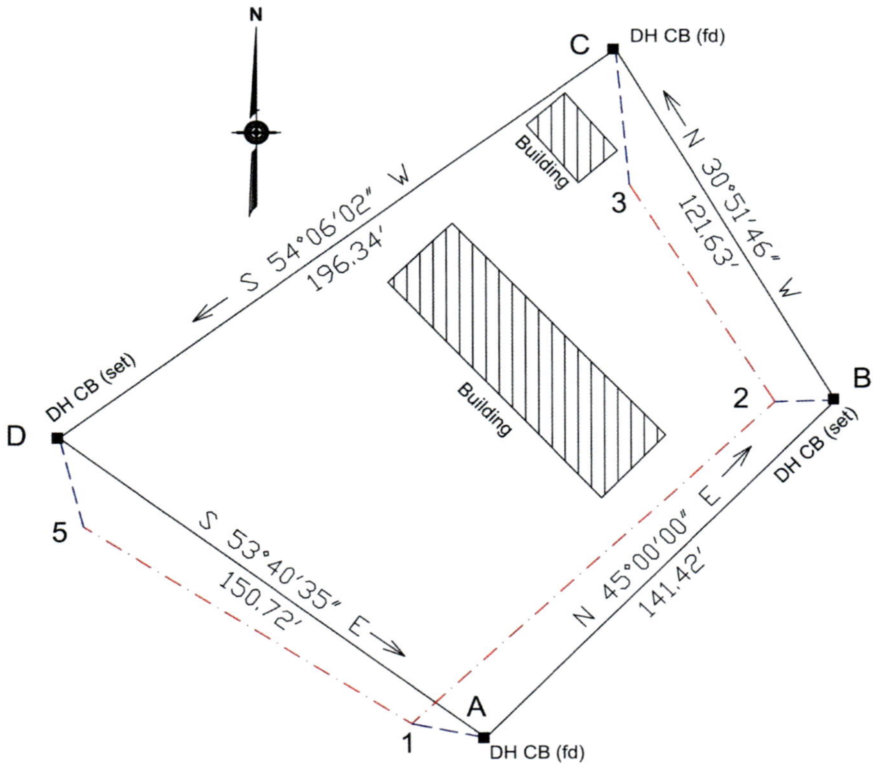

Fig. 12.2 An open traverse

so the survey could have been accomplished without point 4. However, because the traverse would not return to point 1, the point of beginning, there would be no way to know if there was an error in the traverse or, even if an error was suspected, where the error may have occurred. If there was an error in the traverse one or more of the lot corners could have been set incorrectly. A traverse of this type is called an **Open Traverse**. It is occasionally tempting to use an open traverse to save time. But because it is difficult to check for errors, it is generally not accepted surveying practice to use an open traverse for boundary surveying. All boundary surveys should use a **Closed Traverse**.

The tie lines from the traverse points to the lot corners are a potential source of error. If a wrong angle is turned from the traverse line or if a wrong distance is measured the lot corner could be set in the wrong location. Because only a single line ties the corner to the traverse it is not possible to mathematically check for errors. Each of the tie lines really represents a small open traverse. These potential problems can be eliminated by taking an additional measurement to each of the lot corners from another traverse point. Figure 12.3 shows an example of double tie lines. This method effectively creates a number of closed figures which would

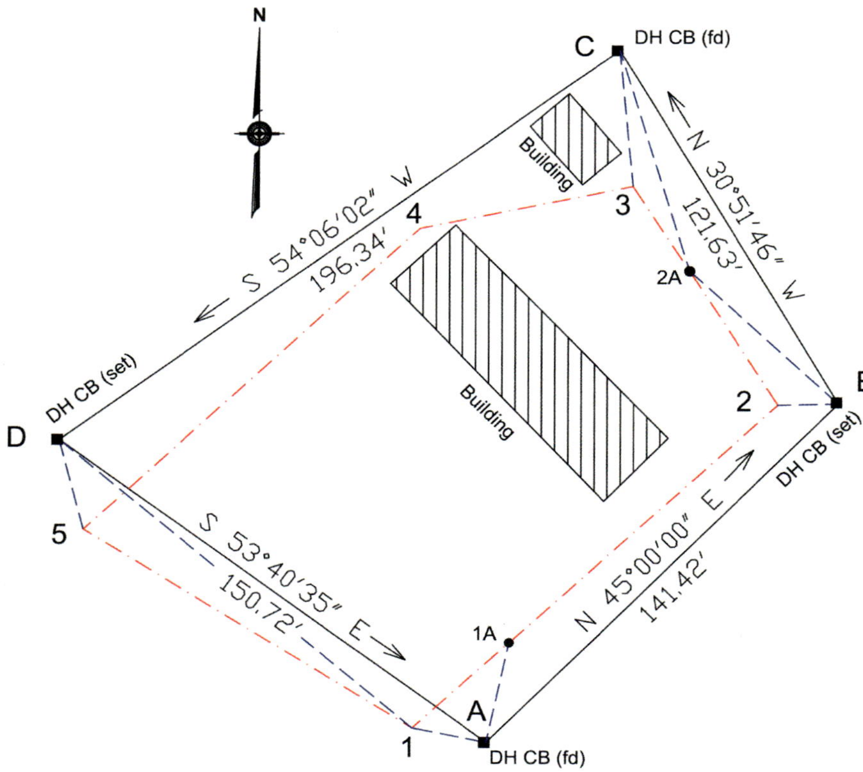

Fig. 12.3 A closed traverse with double tie lines

alert the surveyor of a measurement error. Notice that the points-on-line 1A and 2A needed to be set so that the double locations could be made.

We have seen that most boundary surveys make use of the traverse. If the parcel of land to be surveyed is wooded, the traverse lines will have to be cut. In many cases it is not necessary to cut trees as it is usually possible to run reasonably long traverse lines between the trees just by cutting brush. This can often be done with nothing more than machetes, pruning clippers or a small gasoline powered rotary brush cutter. In the woods, the traverse will need be located reasonably near the property lines because this where boundary evidence is likely to be found. If the traverse lines are too far from the boundary, each measurement to the boundary will require cutting from the traverse line to the boundary. If the boundary is a stone wall or fence and the boundary is very crooked it may be necessary to make measurements to the wall or fence from the traverse line every 10 or 20 feet. In such cases it is far more efficient if the traverse line can be run close to the wall or fence. Because there are often trees or brush between the fence/wall to be located and the traverse line, making it impossible to see the points from the total station, it is often convenient to use a tape to measure offsets from the traverse line to the fence or wall. The offsets are measured at 90° to the traverse line.

As the traverse proceeds along the perimeter of the property it is usual to make measurements to monuments or other evidence from points on the traverse lines where the instrument is located. If a monument cannot be seen from the instrument location it will be necessary to place intermediate points on the traverse lines. These are called "points-on-line". The instrument can then be set on the point-on-line where measurements can be made to the monument. Points-on-line are shown in Fig. 12.3 as 1A and 2A.

Once the traverse has been closed in the field and all of the monuments and other evidence have been located, the data can be brought into the office where calculations are made to verify the location of monuments and to determine how well the actual location of monuments agrees with their record location.

12.1 Using State Plane Coordinate Systems with GPS

The increased use of GPS has made it easier and more common to tie boundary surveys into the state plane coordinate system. If the area where the traverse is to be run is wooded or has other obstructions which will make GPS unreliable, it will be necessary to combine conventional traverse techniques using a total station with GPS control. We can use GPS to establish control points then use a total station to integrate these points into the traverse.

> The increased use of GPS mas made it easier and more common to tie boundary surveys into the state plane coordinate system.

Differential GPS requires one of the receivers to be located at a known point. This point is normally a NGS point for which state plane coordinates are available. Connecting a survey to a state plane coordinate system will provide a number of benefits. There will be additional redundancy because a lost point can always be replaced using its state plane coordinates. For most boundary surveyors, who usually work in a specific geographic area, connecting each survey to the state plane coordinate system will place all of the surveys on a common reference system. As time goes by and physical evidence disappears, having state plane coordinates for this evidence may prove to be invaluable in replacing the lost points. In the past, ties were sometimes relied upon to help recover points should they become destroyed. Unfortunately, ties are also destroyed over time. When traversing, it is possible for a systematic error such as a scale factor, to occur which would not be picked up during the closure and adjustment. If GPS is used to set a number of control points which the traverse then ties into, such errors would be discovered.

GPS can also benefit from conventional surveying techniques. If a survey is to make use of a conventional traverse, the total station can occupy the GPS points as a part of a closed traverse thereby confirming their accuracy. In some cases, the coordinates for a GPS point may not be accurate, particularly if there are obstructions nearby which cause multipath. It is possible that errors in the GPS locations might not be discovered without linear measurements to other known points. There are a number of methods that can be used to confirm GPS locations. One method is to take GPS shots at a location next to a control point. A tape can then be used to measure between the points to confirm their accuracy. Shots can be taken on the control point at different times of the day or on different days. This will allow the collection of data from different satellite geometries. These methods will require additional time and expense. If a traverse is to be run anyway, it makes sense to use total station measurements to confirm GPS locations. In practice, the hybrid surveying technique just described has proven to be both efficient and accurate for boundary surveys.

Connecting surveys to a state plane system raises certain issues relating to the differences in distances between points located by GPS on the state plane coordinate system and those measured on the ground using a total station. In most cases, the distances will not agree. This is not an indication that either the GPS or traverse locations are wrong. The problem is caused by inherent differences between making measurements on the surface of the earth and making measurements on a state plane coordinate system. It is important to understand why this occurs and what steps need to be taken so that boundary measurements are correct. Before GPS became widely used, these issues rarely came up because most boundary surveys were made using total stations for ground measurements and the surveys were not tied into state plane systems. For some surveyors, the use of GPS has required them to work with an unfamiliar state plane system for the first time. While it is undoubtedly true that all surveyors have heard of state plane coordinate systems, not all have actually worked extensively with them.

Connecting surveys to a state plane system raises certain issues relating to the differences in distances between points located by GPS on the state plane coordinate system and those measured on the ground using a total station. In most cases, the distances will not agree.

The reader will recall from our discussion of grid north, in the section on bearing systems, that a state plane coordinate system is a two dimensional map developed from the ellipsoid (earth). The shape of an ellipsoid is simply an ellipse rotated around its principal axis. The earth has a slightly larger diameter at the equator than at the poles so it is technically not a sphere—although it is nearly so. The ellipsoid is approximately mean sea level. The ellipsoid is also called a spheroid. When we refer to a state plane coordinate system we are referring to a two dimensional plane coordinate grid which slices through the earth much like a knife slices through a grapefruit as in Fig. 12.4. Imagine that there are points on the upper surface of the grapefruit which must be projected onto the flat surface of the knife blade.

When the ellipsoid shape of the earth is projected onto a flat surface, distortion is inevitable. The amount of distortion can be minimized if the size of the projection (Zone) is kept relatively small. If the zone is kept to a maximum size of 158 miles, the distortion is kept to 1 in 10,000 or less. Recall that 1 in 10,000 would be a reasonable closure precision for a boundary survey traverse. The size limitation of the zone means that large states such as California and Texas must have several

Fig. 12.4 A knife slicing a grapefruit similar to a state plane grid slicing the earth

zones, while small states such as New Hampshire, Connecticut and Delaware are able to have a single zone.

The distortion from projecting points from an ellipsoid to the state plane surface is shown in Fig. 12.5. This figure is greatly exaggerated in order to show the distortion clearly. The state plane projection shown in Fig. 12.5 can be visualized as a plane slicing through the earth—like our knife slicing through the grapefruit. In the image, the line representing the state plane grid is as if we were looking at the sharp edge of the knife blade. In the example, projections from the ellipsoid to the plane surface are done radially from the center of the earth. Notice that the radial lines are not parallel to each other but become wider apart as the distance from the center of the earth increases. Because the ellipsoid surface and state plane surface are separated, the distance between any pair of radial lines will be different at each surface.

The central meridian is a line that represents true north. It can be seen that the state plane surface cuts through the earth such that the center portion is below the upper ellipsoid surface and the outer portion is above the ellipsoid surface. The state plane surface is called a secant, which is simply a straight line intersecting a curve at two points. The projection on the left, which is inside the ellipsoid, causes the distance on the state plane surface to be less than it is on the ellipsoid. The ellipsoid distance is 530.00 and the projected distance is 525.00. The opposite is true when the plane surface is outside the ellipsoid, as in the projection at the right, where the ellipsoid distance is 520.00 while the projected distance is 525.00. When using the small zones typical in actual state plane coordinate systems, these differences would normally be on the order of a few tenths or hundredths of a foot, not many feet as in our exaggerated example.

Notice in Fig. 12.5 that there are two places where the state plane crosses the ellipsoid surface. These are the only two places in a state plane zone where the state plane dimensions will be equal to the ellipsoid dimensions. In other words, the distortion is zero. Also notice that where the state plane grid is below the ellipsoid surface, the amount of distortion decreases as the distance away from the central meridian increases. This makes sense because the distortion becomes zero where the state plane grid crosses the ellipsoid surface. Notice also that after crossing the ellipsoid surface, the distortion begins to increase as the distance becomes further from the central meridian. It is possible to develop a mathematical relationship which will calculate the amount of distortion on the state plain projection as a function of the distance from the central meridian. In other words, if we know how far our survey is from the central meridian, we can calculate precisely how much our distances will be shortened or lengthened relative to the ellipsoid distances.

Boundary surveying usually does not take place on the ellipsoid, but at some ground elevation above it. So we must also consider the additional distortion caused by the elevation of the earth's surface above the state plane level. A hypothetical ground surface is shown in Fig. 12.6. If we look at the projection on the left, we can see that the state plane distance is 525.00 and the distance along the surface of the ground is 535.00. It is apparent from Fig. 12.6 that the elevation of

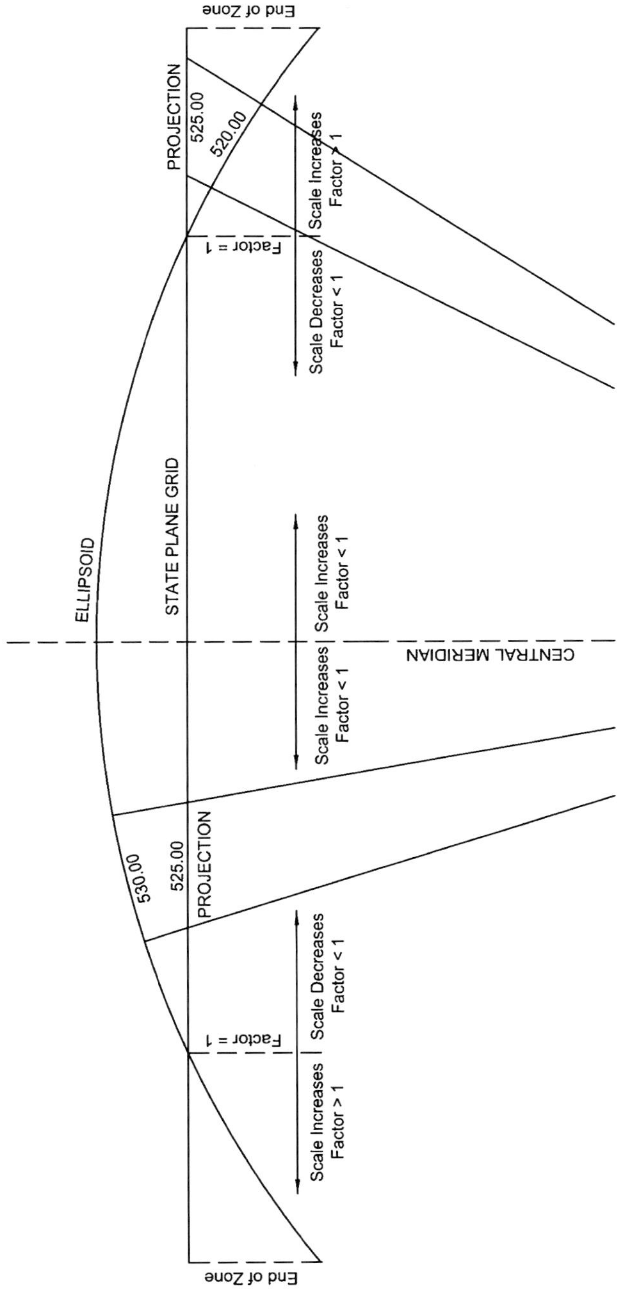

Fig. 12.5 State plane projection distortion

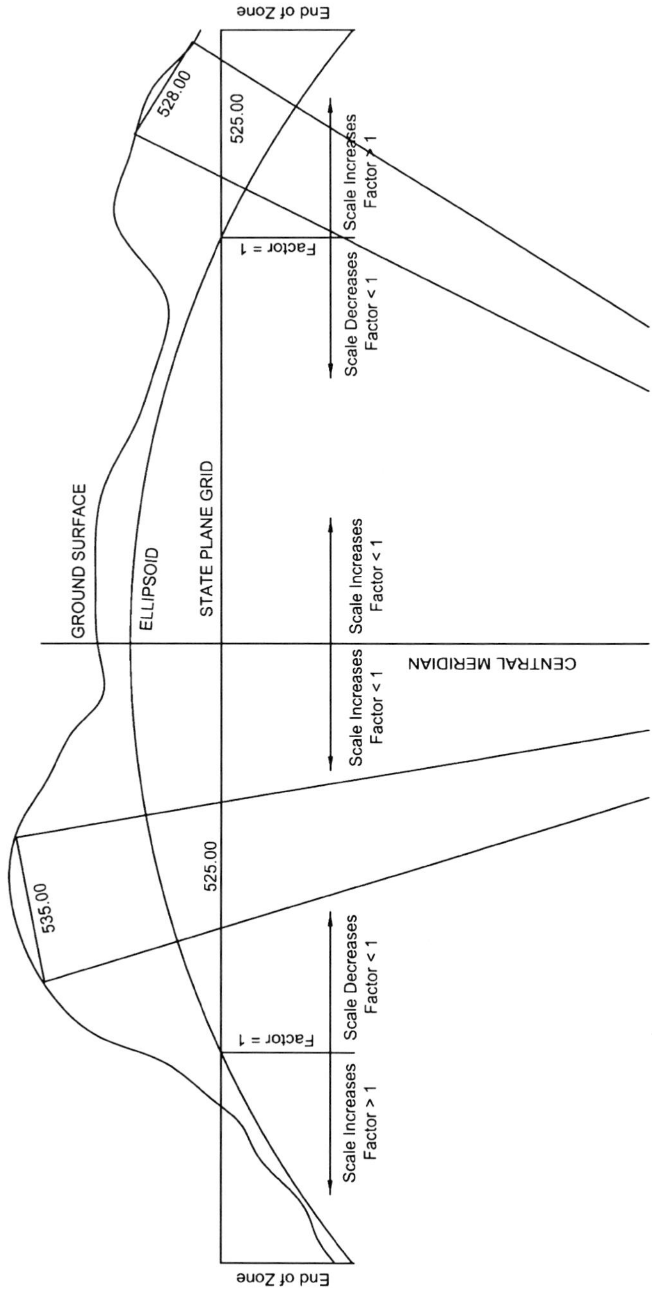

Fig. 12.6 State plane projection ground

the ground has an effect on the distortion, and the higher the elevation, the greater the distortion. In fact, because the zone size of a state plane system is limited, in mountainous areas the elevation of the ground can have a greater effect on distortion than the state plane/ellipsoid distortion. It is important to understand that when we speak of distortion we are talking about distances which are measured on the ground between points. Distances measured on the flat state plane grid will be different than distances measured along the surface of the earth. This is the primary issue faced by surveyors when working with state plane coordinates.

Based on Figs. 12.5 and 12.6, we can conclude that in order to convert from state plane distances to ground distances, we need to know how far the points are from the central meridian and we also need to know the elevation of the ground where we are performing our survey. In practice, a factor called a grid factor, also known as a combined grid factor or combined factor is used to convert ground dimensions to grid dimensions and the other way around. A combined factor is a number that can be multiplied by a distance in one system in order to convert the distance to the other system. The combined factor takes into consideration the distortion between the ellipsoid and the state plane grid and the elevation of the ground where the survey is located. At the point where the state plane grid crosses the ellipsoid surface, the combined factor will be 1 (for an elevation of zero). As we already discussed, there is no distortion at this point. At other points on the state plane grid, the combined factors will typically be close to 1, such as 0.999777 or 1.000215. We can see from Fig. 12.6 that, from the central meridian to the point where the state plane grid crosses the ellipsoid, the combined factor is less than one and it increases as the distance increased from the central meridian.

If we multiply the combined factor times the ground distance, we will have the distance on the state plane grid. For example, if we measure a survey line on the ground that is 1,000.00 feet long and we have a scale factor of 0.999777 the distance on the state plane grid would be 999.78. In our example, the difference is 0.22 feet. If our survey were a small house lot with a boundary length of 100.00 feet the distance on the state plane grid would be 99.98 feet, a difference of only 0.02 feet. As our example demonstrates, unless the survey is very large, or if it is at a very high elevation, the difference between ground distances and state plane distances will not be large, but the differences will still be enough that they will need to be corrected—at least when the survey requires high accuracy, such as when establishing boundaries. For survey work in which accuracy standards are less demanding, correction may be less important. Surveying software running on a personal computer or in a data collector will usually have scale factor conversion built in so that a user can easily switch from one coordinate system to the other.

When working with conversions between ground coordinates and state plane coordinates, it is important to realize that the coordinates themselves are not distorted. If we have a grid of one foot squares on the state plane grid and a grid of one foot squares on the ground surface where we are performing a boundary survey, the squares on both surfaces will measure exactly one foot. However, the distances between points projected from the ground grid to the state plane grid will have different lengths. Mistakes have been made in the past by applying scale factor corrections to coordinates. Combined factors are only applied to distances. Coordinates are not scaled. When we speak of converting coordinates from one system to the other we are just saying that the values of the coordinates are changed. We do not change the length of the foot or meter which is used for measurements of or on the grid.

When working with conversions between ground coordinates and state plane coordinates, it is important to realize that the coordinates themselves are not distorted.

In boundary surveying, we are primarily interested in the distance measured on the ground. In most cases the distance will be measured with a total station. As an example, suppose we begin a survey by using GPS to set a number of control points on a 200 acre boundary survey. We will perform the remainder of the survey with a total station and use the GPS control points as traverse points. We decide that the GPS coordinates will be based on the state plane coordinate system. Let us also assume we are in an area where the ground elevation and distance from the central meridian causes a substantial amount of distortion. When we go out into the field and measure between the GPS points with our total station, the distances which we measure will not agree with the distances calculated from the GPS measurements. As we have seen, for very short lines, the differences may not be that large, but for long lines, the differences may be substantial.

When working on a survey which incorporates state plane values, a surveyor must decide which coordinate system to use. He can use the state plane values or he can use ground values. The calculations can proceed either way, by simply converting coordinates from one system to the other. Because GPS is now being used in many boundary surveys, it would seem to make sense to work on the state plane coordinate system. Should a surveyor decide to convert the GPS control to a ground based coordinate system, and to work on that system it should be kept in mind that a single scale factor will only be accurate for a small survey. So, if the survey covers a large area it will be more accurate and less confusing, to work with state plane coordinates, otherwise more than one scale factor may be needed, making the calculations more cumbersome.

If a surveyor decides to work on the state plane system, boundary dimensions will have to be converted to ground dimensions so that that the distances between lot corners are correct when they are measured on the ground. The decision to use a state plane coordinate system does not convey a license to the surveyor to modify boundary dimensions from their true values. Boundary surveys often require that plans be prepared. Whenever a survey plan shows boundary dimensions there are legal implications. For example, when a surveyor prepares a plan to be used as the basis of a parcel description, the distances on that plan must be actual distances as measured on the ground, not distances adjusted to fit the state plane coordinate system. When a client purchases a parcel 1,000.00 feet square she does not bargain for a parcel 999.60 feet square. Nor does she bargain for less area. When a surveyor is retained to resurvey a parcel of land created in 1925, he must follow in the footsteps of the original surveyor. He does not have the authority to change the size of the footsteps.

Some jurisdictions require certain surveys to be connected to state plane coordinates. Many state statutes also provide that state plane coordinates are an acceptable means of locating a boundary corner and that said coordinates constitute a legal and satisfactory description of the corner. These statutes usually contain a provision that no purchaser or mortgagee of real property is required to depend solely on state plane coordinates to describe a corner. A purchaser may still require physical monuments—and they should. When ground values are shown on a plan and the survey is tied into a state plane coordinate system, the surveyor must be careful to show enough information so that anyone interpreting the plan will know how to reproduce the survey on the ground and to connect it to state plane. This must include clear descriptions and identifications of control points, basis of the coordinate system and units, scaling factors and ties from the control to the parcel being described.

12.2 Low Distortion Projections

Low Distortion Projections (LDP) are becoming a common alternative to state plane coordinates, particularly in mountainous areas having high elevations which cause substantial distortion. An LDP is simply a coordinate grid which is located at a height above the ellipsoid which is the approximate average or median elevation of the topography where the survey will take place.

A local central meridian is selected at the representative ground surface height as shown in Fig. 12.7. To minimize distortion, the coverage of the LDP is designed to only cover the area of interest. The low distortion projection plane shown in Fig. 12.7 is midway between the lowest and highest elevation (which may not be the average elevation). The custom scaling factor would be valid for the darkly hatched area. The LDP allows a single scaling factor to be used for the entire project. The LDP has the advantage of reducing the differences in distances measured with GPS and total stations. This feature is a great advantage over state plane coordinate systems, particularly when working in areas with high elevations. Some surveyors believe that with the high accuracy possible with GNSS, and the issues between ground measurements versus state plane, the state plane systems are becoming obsolete and LDP may offer a better solution. Many data collectors and surveying calculation programs now allow a user to create custom LDPs with conversions to state plane coordinates. A number of states are considering offering multiple LDPs for their states. Until LDPs become a part of the legal framework of a jurisdiction, there is no official basis for a particular LDP. In order to keep distortion to a minimum the LDP area must be smaller than a state plane system so many LDPs will be required for an area the size of a state.

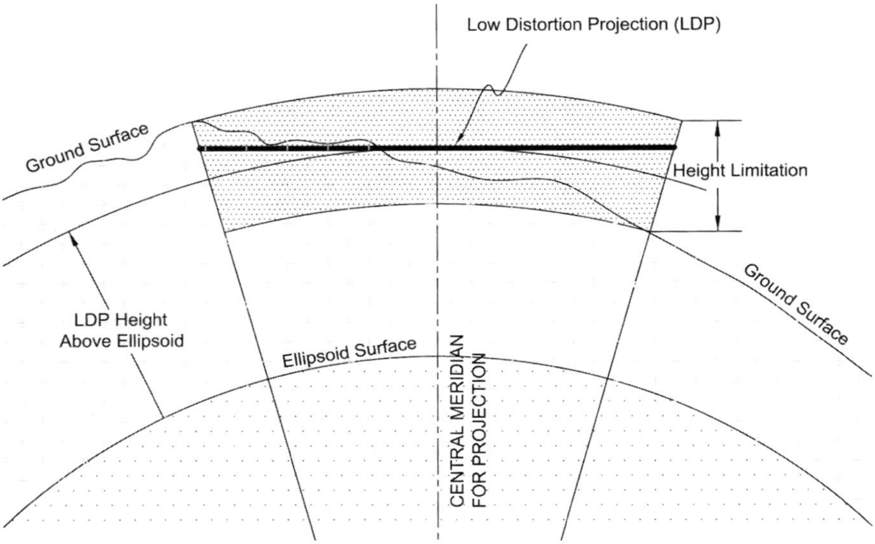

Fig. 12.7 Low distortion projection

12.3 Using GPS in Boundary Surveying

Let us look at an example of a static GPS survey to see how a simple GPS network can be used to connect a boundary survey to a state plane coordinate system.

A simple GPS network is shown in Fig. 12.8. The network consists of three points: NGS1, C1 and C2. The property to be surveyed is labeled Locus. The survey of the Locus will be made using a total station by running a traverse around the inside perimeter of the property. The property owner desires that the property survey be tied into the state plane coordinate system. The area is completely wooded so it is not possible to see between C1 and NGS1 or C2 and NGS1 (NGS1 may be several miles away from C1 and C2.). Because C1 and C2 are on the edge of Holly Way, it is possible to see along the highway between the two points. In fact, the locations of C1 and C2 were specifically chosen to be intervisible.

NGS1 is an NGS control point which has known coordinates on the state plane coordinate system. C1 and C2 are control points that have been set at randomly picked locations. When selecting locations for GPS control, the surveyor must carefully plan in advance how the project will proceed. Our example is a simple one but many projects will include many points in the control network. In this example we will need to be sure that there is an open view of the sky from C1, C2 and NGS1 so that as many GPS satellites as possible are visible at one time. As we have already seen, there should be no large buildings in the vicinity which might cause multipath problems. The control points should also be intervisible—it should be possible to see from C1 to C2 so that we can use a total station to measure between these points. The line from C1 to C2 will become a traverse line

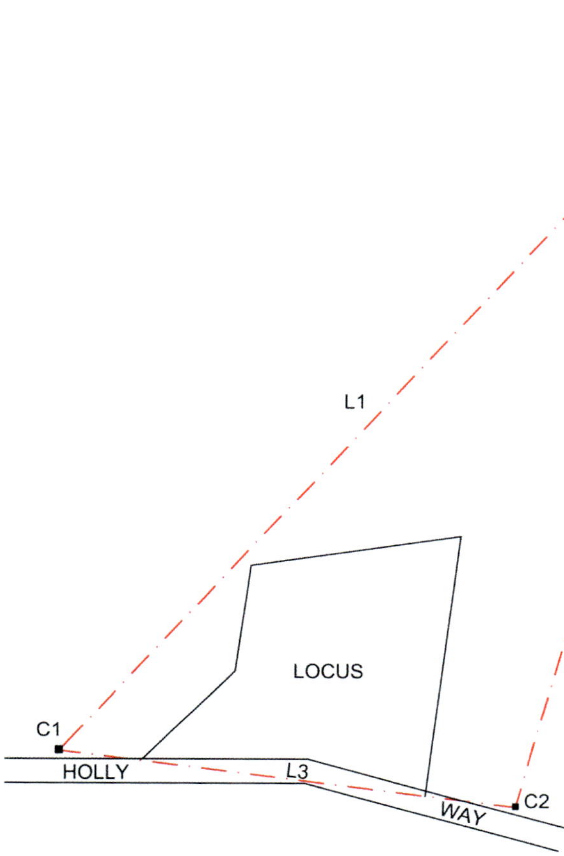

Fig. 12.8 GPS control network

which will be used to run a traverse around the interior of the Locus. The traverse consists of points T1 through T4 as shown in Fig. 12.9. It can be seen that the points T1 and T4 are intermediate points on the straight line between C1 and C2. By including line C1–C2 in the traverse, the traverse will automatically be tied into the state plane coordinate system. As we noted, the locus is completely wooded so lines will need to be cut through the woods between T1, T2, T3 and T4 so that we can use our total station for the traverse part of the work.

For this example, we will assume that only two GPS receivers are available. With static GPS it is necessary to collect data for each line in the control network. Additionally, data is only used for the period when both receivers are simultaneously collecting data. In our example, during the planning phase of our survey we learned the best time of day to make the observations and determined that the receivers needed to collect data for at least 1 hour. To begin the GPS portion of our survey, one GPS receiver will be placed on C1 and the second receiver on NGS1. If we only

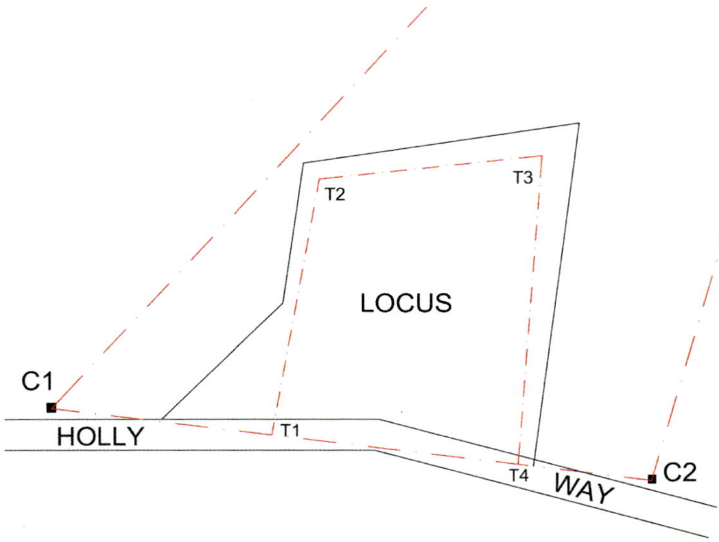

Fig. 12.9 Hybrid GPS and conventional survey

have one person in the field, it will be necessary to set up the receivers on points C1 and NGS1, turn on receiver C1, drive to NGS1 and turn on that receiver. If it took 15 min to drive from C1 to NGS1, that portion of the data would not be used because both receivers would not have been on for that period. The same would be true for turning off the receivers. The only data that would be useable would be the data collected while both receivers were simultaneously collecting data.

Once we have collected data for the required amount of time, we would have enough data to calculate survey line L1. We would then move the C1 receiver to C2 and collect data for another hour. This will give us data for survey line L2. Finally, we would move the receiver at NGS1 to C1 and repeat the process for survey line L3. In our example, it would take at least 3 hours to collect the data for our network. If we only had a single person working it would take longer, depending on the time required to travel between the points. Notice that we have to collect data for each line in our network. It can be seen that if we had three receivers instead of only two, data for all points would be collected simultaneously and it would only take us an hour to collect all of the data, instead of 3 hours.

A word of caution is in order when leaving GPS receivers in the field. There is an understandable tendency to leave the receivers unattended when they are collecting data. Doing so reduces the size of the field crew. It also seems like a waste of time to have a crew member stand there for an hour or more watching the receiver. However, leaving the receivers unattended may be too great a temptation for some and having the receivers stolen or even mischievously moved or tipped over is not unheard of. Because of the necessity of simultaneous data collection, loss or movement of a receiver means starting over again for that phase of the work.

Once we have collected all of the data, we would bring the receivers into the office and upload the data to a personal computer where the network would be analyzed and adjusted. The requisite software is usually supplied by the receiver manufacturer. When the data is analyzed, we would instruct the software to hold point NGS1 and enter the state plane coordinates for that point. Because GPS collects 3-dimensional data, we would also need to enter the elevation of the point. Assuming there were no flaws in the data, after adjusting the network the software would calculate coordinates for points C1 and C2. The bearings and distances for the three lines would also be calculated. The coordinates would then be uploaded into our total station data collector. When we went back into the field to run the traverse around our Locus, we would begin by setting the total station on point C1. After setting a prism on C2 we would use the total station to measure the distance between C1 and C2. Because our GPS software has already calculated this distance, we would be able to compare the results of the GPS measurement and the total station measurements. This would provide us with a confirmation that this portion of our network was accurate. Of course we would need to take into consideration the differences between ground and state plane grid measurements that we discussed earlier. While the total station occupied C1 we would place points T1 and T4 online so we could begin our traverse through the wooded parcel.

As our example demonstrates, GPS can be a valuable asset for a boundary surveyor. For large boundary surveys a number of control points can be located throughout the area where traverse lines will be run and where the total station will be able to occupy these points as a part of the traverse. This will provide redundant control and the traverse can be adjusted between the GPS points.

Static GPS allows even small boundary surveys to be economically and quickly tied into the state plane coordinate system. Once a survey is tied into state plane, it will always be possible to reproduce the survey, even if all of the permanent monuments on the site are destroyed. Static GPS can also be useful when it is necessary to locate evidence that is too far away to be located economically using conventional traversing techniques. One example that comes to mind is when a state, county or town line runs through the property being surveyed or the line constitutes one of the parcel's boundaries, and the closest monuments on the line are miles away. Traversing to the monuments may take days but, using GPS it might be accomplished in a few hours.

Consider the survey shown in Fig. 12.10. A state line runs through the locus. The locus traverse consists of points 14 through 17. A state line boundary marker, SB1, was found in a wooded area two miles away. A GPS receiver cannot be set on the boundary marker because it is in a wooded area surrounded by tall trees. One way to locate the boundary marker would be to establish a random short line just outside the wooded area in the open. This line is labeled 1–2. GPS receivers can then occupy points 1 and 2 and one or more of the points on the locus traverse. The points on the locus traverse would serve as control. Once coordinates for points 1 and 2 have been obtained, a total station could be set up on point 2, a prism set up on point 1 and SB1 could be located using an angle and distance. In this example it would not be necessary to traverse two miles and back in order to

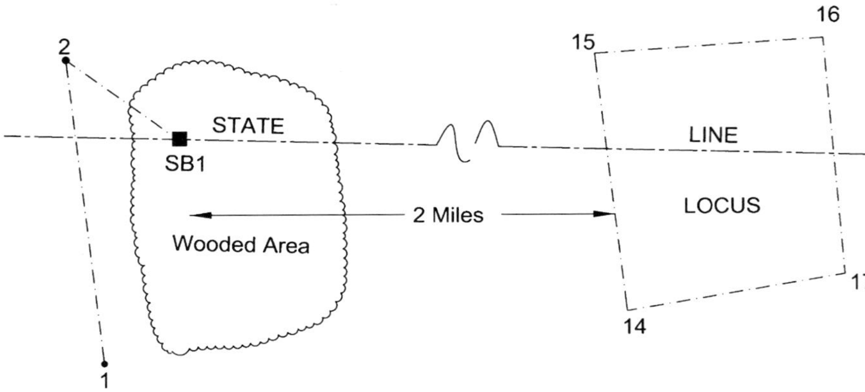

Fig. 12.10 Static GPS remote location

locate SB1, which might take days. The GPS location could probably be accomplished in a much shorter time. Of course, in order to complete the survey and determine the direction in which the state line runs, it would be necessary to tie into an additional monument somewhere on the line, preferably on the right side of the locus. An additional consideration is that it may be preferable to set three points with GPS near the wooded area so that their accuracy could be checked with the total station. Although their absolute positional accuracy would not be confirmed, their relative accuracy could be confirmed. Occupying additional points at the locus with GPS would be prudent as it would strengthen the network.

Chapter 13
Water Boundaries

In many cases a surveyor must establish the boundaries of a parcel of land bordering a river, stream, lake or the sea. If the boundary is a river or stream, the rights that a person has are referred to as **Riparian Rights**. If the boundary is an ocean or a lake, then the rights are called **Littoral Rights**. Sometimes the rights relating to any water boundary are referred to as riparian rights but this terminology is inaccurate. The law of water boundaries varies considerably between jurisdictions. What follows is a general discussion designed to acquaint the reader with some of the issues related to water boundaries in some jurisdictions. The reader is encouraged to research the law in their own jurisdiction.

The first step in locating the boundaries of a parcel of land fronting on the water is to determine if the water is navigable. In general, the owner of land abutting non-navigable waters owns the bed. The bed is the area between the banks that is normally covered by water. If the water is navigable, the property owner will usually not own the bed.

The question of whether a particular body of water is navigable is not always clear and the test for navigability varies between jurisdictions. The federal rule for determining whether a body of water is navigable is whether or not the water is **Navigable-in-fact**. The test is whether the water is used, or is susceptible of being used in its ordinary condition as a highway over which commerce may be carried on between other states or countries. If the water is navigable, then ownership of the bed rests with the state. If the water is regularly used for commerce, there should be little question as to the upland owner's rights, the ownership of the bed will be vested in the state. If the water is not used for commerce and it is not capable of being used for commerce, ownership of the bed will most likely be with the upland land owner.

Where a stream is non-navigable, the upland owner bordering on one side of the stream is presumed to own to the middle or **Thread** of the stream. The thread is a line that is midway between the banks of the stream, usually measured at

P. Gay, *Practical Boundary Surveying*, DOI 10.1007/978-3-319-07158-9_13

Fig. 13.1 Property bordering a stream is presumed to go to the center or thread of the stream

average high water (Fig. 13.1). The presumption of owning to the stream thread applies if the deed merely calls for the stream in general terms, as a boundary. For example if a deed said that the property ran along Whiskey Creek or Angela Brook, the owner would be presumed to own to the thread of the stream. A deed may have words of exclusion, providing that the ownership only runs to the edge of the stream. For example, if a deed says that the line runs to the high water mark of Gold Creek, the upland owner would not own the bed of the stream. The thread of a stream should not be confused with the channel or deepest part of a stream, sometimes called the **Thalweg**.

> Where a stream is non-navigable, the upland owner bordering on one side of the stream is presumed to own to the middle or Thread of the stream.

Along navigable tidal waters such as the ocean shore or tidal rivers, the upland owner does not own the bed. Under federal law, ownership extends only to **Mean High Water**. Mean high water is determined by calculating average tidal heights for a period of 18.6 years. Although the tide is affected by both the sun and the moon, the moon has the greatest effect. The 18.6 year period is the time required for the declination of the moon to complete one cycle. The Federal Government maintains

Fig. 13.2 Under Federal law, private ownership of land bordering navigable water runs to high water, but there are state exceptions

tidal measurement stations along the U.S. coast and tidal datums are available to surveyors. When the tide moves into shallow water estuaries, tidal heights can be substantially affected. A tidal datum established for the mouth of a river may not be applicable for other points further upstream. Because there may not be tidal data available for many locations in shallow water estuaries it may not be possible to accurately determine the height of mean high water at these locations without an actual tidal study. Such a study would probably be considerably beyond the scope of the average boundary survey. Often, the slope of the shoreline in tidal waters is very shallow so that small variations in tidal elevations have the potential to cause substantial differences in the geographic location of the water boundary (Fig. 13.2).

Some states do not follow the federal rule and do not use mean high water as the limit of private ownership. California uses neap high tide which is the high tide level corresponding to the period when the sun and moon are in quadrature (the sun and moon are about 90° apart as measured from the earth). In another example, in Massachusetts, a Colonial Ordinance established that an upland owner owns the bed to low water but no more than 100 rods (1,650 feet) from high water. New Hampshire and Maine follow the same rule. There are also jurisdictional differences in the ownership of the beds of fresh water ponds and lakes. In Massachusetts, the beds of ponds over 10 acres, called great ponds, are owned by the Commonwealth.

Water boundaries are not static and are subject to continuous change. Streams meander, rivers change their boundaries and ocean shorelines change their shape and location. There has been much litigation over how these changes affect littoral or riparian rights. Here again, there are differences between the states.

Accretion refers to the gradual accumulation of land by natural causes along a water boundary. **Reliction** is the increase in land caused by the permanent withdrawal of water. According to common law, soil added by accretion and reliction becomes the property of the upland owner. Where the boundary is the thread of a stream and the stream gradually changes direction and moves away from the upland, thereby adding land, the upland owner continues to own to the thread of the stream. The added land belongs to the upland owner. Where a stream moves further onto the upland through the gradual process of erosion, the upland owner loses the land. Reliction also occurs when lake waters subside.

Avulsion means the sudden and perceptible removal or addition of land by action of water or by a change in the course of a stream. Where a stream suddenly changes direction the boundaries do not change. The boundary location remains at the location of the original thread of the stream even though the stream has suddenly relocated itself. Where a stream forms a new course, an upland owner who originally owned property only to the thread of the stream, may then own on both sides of the stream, while the owner originally on the other side of the stream will no longer own to the stream.

13.1 Tie Lines

As a result of the irregular nature of water boundaries, and because of the fact that water boundaries are often determined by water elevation, it is often difficult to provide precise dimensions to, or along these boundaries. When a plan is drawn showing accurate bearings and distances around the perimeter of the property, the lack of accurate dimensions along the water boundary does not allow a precise boundary closure for the perimeter of the parcel. Surveyors get around this problem by using one or more **Tie Lines**. A tie line is simply a line that ignores the water boundary. It is a line that connects the accurately dimensioned boundaries, providing the missing bearing and distance so that the perimeter will have a mathematical closure. See Fig. 13.3 for an example of a tie line.

A tie line is simply a line that ignores the water boundary. It is a line that connects the accurately dimensioned boundaries, providing the missing bearing and distance so that the perimeter will have a mathematical closure.

A tie line has nothing to do with limits of ownership. It is merely a mathematical convenience. It provides evidence that the description of the property boundaries is mathematically correct. To avoid confusion about the actual claim of ownership,

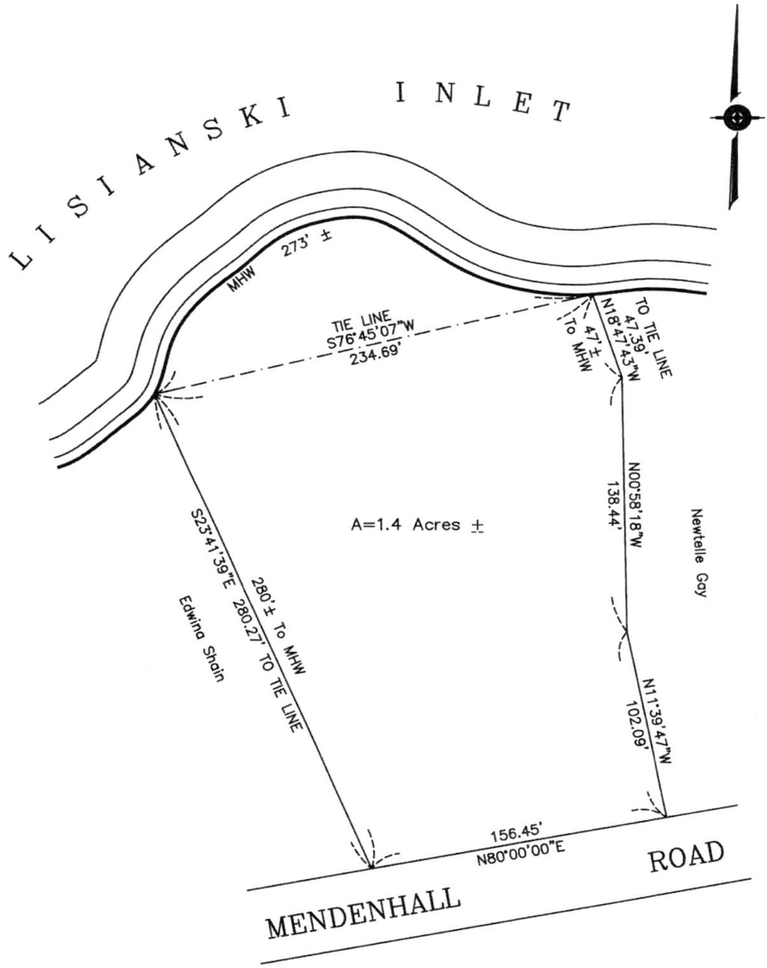

Fig. 13.3 Tie line along water boundary

tie lines should not be recited in deeds; rather, the description should state the dimensions along the water boundary even if they are termed "more or less". The deed should call for the plan which shows the tie line. As the plan will then be incorporated into the deed, both the attorney and the surveyor will have the information they need.

Chapter 14
Calculations

In the calculation phase of a boundary survey, the evidence and information gathered during field work is brought into the office. Mathematical adjustments can be made to the traverse and the physical evidence found in the field can be compared to the record information in order to determine the extent of agreement or disagreement. This is the phase of the survey where a surveyor will make a determination as to the probable location of the boundaries of the parcel being surveyed.

14.1 Coordinate System Example

As we have seen, when using a total station to make locations in the field, the result is a series of angles and distances. These are written in a field book or stored in a data collector as a raw file. We will generally use the raw data to create a coordinate file. This will normally be done on an office computer.

Since we are primarily interested in boundary surveying, and all dimensions are horizontal, we will be working on a two dimensional plane. Recall that surveys are usually plotted using a Cartesian coordinate system. Values increase up and to the right. The axes are labeled **Latitude** and **Departure**. Latitude (Y) is measured north and south and Departure (X) is measured east and west.

In Fig. 14.1 the grid represents our coordinate system. For the small boundary survey pictured, we did not tie into a state plane coordinate system, so we will use magnetic north as the basis for orienting our grid. Because our coordinate system is assumed, the absolute value of the coordinates does not matter. Considering the maximum size of our survey, we simply need to pick a value for our starting coordinates large enough that they will not become negative. There is nothing wrong

P. Gay, *Practical Boundary Surveying*, DOI 10.1007/978-3-319-07158-9_14

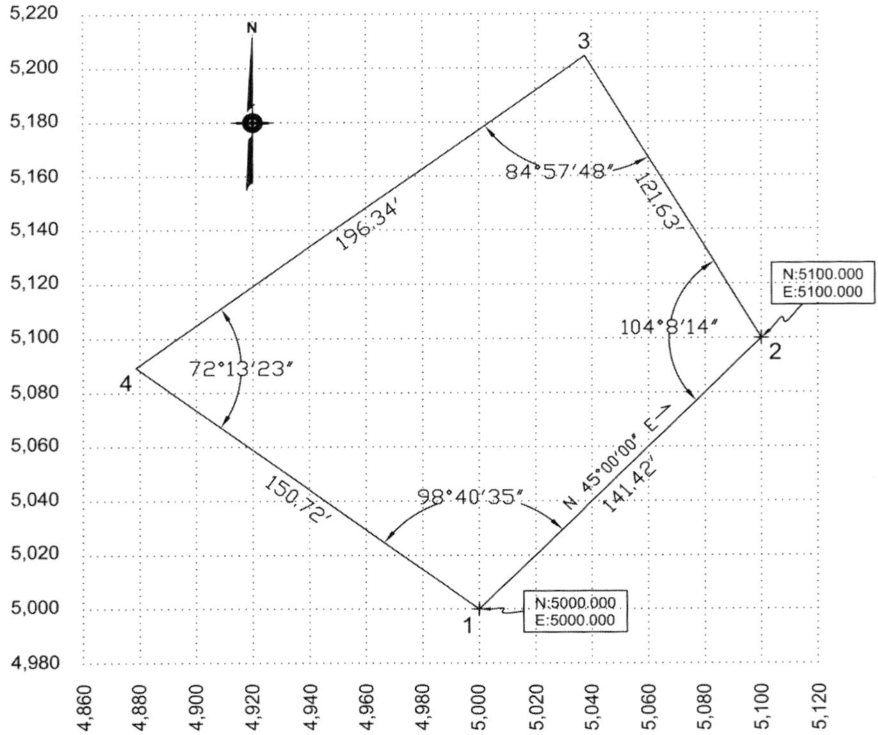

Fig. 14.1 Angles and distances converted to coordinates

with negative numbers, but it is easier just to pick relatively large numbers such as N5,000, E5,000 so we don't have to deal with minus signs.

We will assign Point 1 the starting value of N5,000, E5,000. During our field work, the line from 1 to 2 was sighted with a magnetic compass. The magnetic bearing was N45°E. By starting at Point 1 and going N45°E, 141.42 feet, and using a little trigonometry, we can calculate the coordinates of Point 2 (N5,100, E5,100). Recall that a bearing in the NE quadrant is simply an angle measured from north in an easterly direction. So our bearing N45°E is simply an angle which is 45° to the right of north.

Because we know the bearing of line 1–2 and the angle to point 3 (104° 08′ 14″), we can calculate the bearing of the line 2–3. The appendix has examples of calculating bearings using angles. Using this bearing and the distance of 121.63 feet we can calculate the coordinates of point 3.

With this method, we can calculate the bearings of each of the lines and the coordinates for each of the points, as shown in Fig. 14.2. This may seem like a lot of effort, but our computer will do most of the work for us. In fact, software is

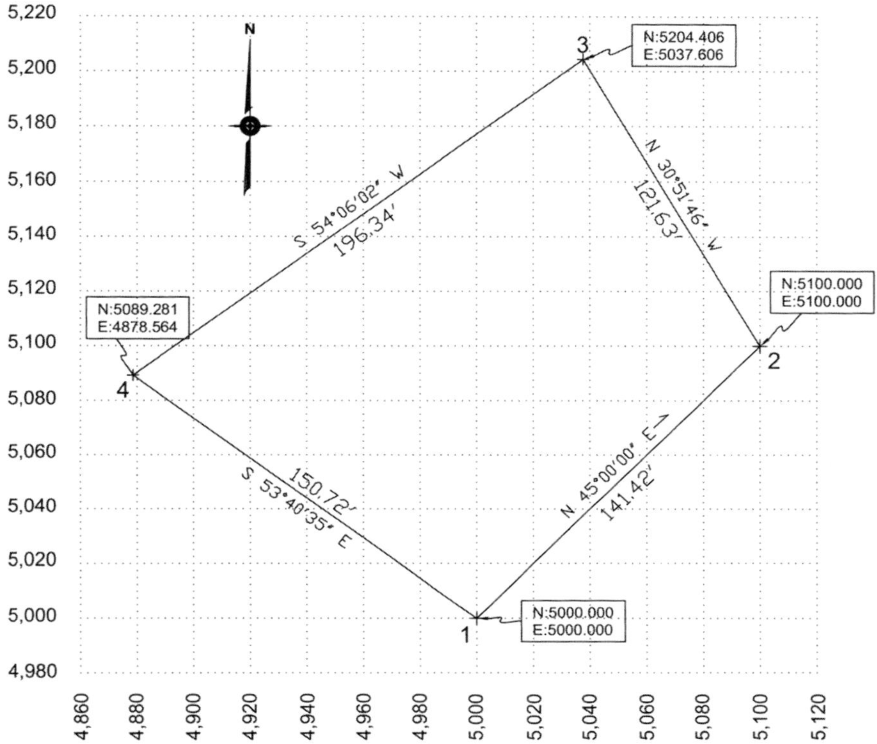

Fig. 14.2 Coordinates shown as Latitudes (*N*) and Departures (*E*)

available which will take our raw data file and, in a second or two, calculate the coordinates of all the points we located.

Coordinates are useful because, once we have coordinates for our locations, it is a simple matter to calculate bearings and distances between them. The process is essentially the above described process in reverse. Being able to calculate the bearing and distance between points will allow us to see how well our physical evidence fits the legal description of the property being surveyed.

14.2 Traverse Adjustment

Before we can reconcile the evidence, we must adjust our traverse to remove any measurement errors. The adjustment procedure distributes the errors in measurements and leaves the surveyor with a traverse that closes perfectly—at least on paper. All field measurement techniques involve some amount of error. If a total station is used to measure traverse lines, there will be errors in sighting the prisms—no matter how careful the instrument person is. A tripod may not be set up exactly over a point, or the tripod may be set up on ground which is soft or thawing, causing the tripod to move slightly between measurements. This would cause errors in both angles and distances. These errors can accumulate, or they can cancel each other out. Usually it is some combination of both. The purpose of adjusting the traverse is to remove these errors.

The reader is already familiar with the concept of a closure from our previous discussion of the property line closure. The same closure theory which we applied to property lines applies to traverses. The surveyor must know how well the traverse closes in order to determine the precision of the measurements that were made in the field. After the traverse is adjusted, most of the work will involve reconciling the record evidence with the physical evidence. In many cases this will be the most difficult and time consuming part of the calculation process.

A detailed discussion of the mathematics used to adjust a traverse is outside the scope of this chapter but examples are provided in the Appendix. A few points should be mentioned. The first point is that the traverse must close within certain limits. These limits may be imposed by law or they may be based on the standards of practice commonly used by surveyors in the area for the type of survey being conducted.

Typically the surveyor will enter the raw traverse data, which consists of angles and distances measured in the field by the instrument, into a computer program. The program will then calculate the error of closure. In actual practice, most surveyors use data collectors on their instruments. These data collectors are small computers that often have the ability to inform the user, at least on an elementary level, approximately how well the traverse will close. This information is usually available to the surveyor while in the field as soon as the traverse returns to the beginning traverse point. Sometimes the data collector is capable of adjusting the traverse. However, even if a data collector is capable of adjusting the traverse, it is often more convenient to use a more powerful office computer that can take advantage of more sophisticated mathematical routines to adjust the traverse.

Once the surveyor has determined that the error of closure is within allowable limits, the traverse is adjusted. There are a number of mathematical methods of adjusting a traverse but suffice it to say that the adjustment takes the closure error and distributes that error at certain places in the traverse so that the adjusted traverse closes perfectly. Some of the more sophisticated closure routines are able to determine the most likely places where the errors occurred and place the necessary adjustments there. Other routines merely distribute the error proportionally over all of the traverse lines. With modern surveying instruments and techniques the errors are usually very small and the closures usually exceed minimum closure requirements by a substantial margin.

After the traverse is adjusted, the locations of all monuments and other evidence found in the field are calculated, and the coordinates of the points plotted so that the surveyor can visualize and calculate the relationships between these points. Most surveyors now use Computer Aided Design (CAD) programs so that the relationships between the points can be determined on a computer screen. If the surveyor has a plotter, a large format printer, the CAD program can send the information to the printer and the surveyor will then have a plan showing the information. CAD programs eliminate the need for the time consuming process of manually plotting each of the points on graph paper. As a large boundary survey may have upward of a thousand points, considerable time can be saved by using CAD.

14.3 Reconciling the Record and Physical Evidence

The surveyor should now have much of the information necessary to calculate boundary locations. If the survey is an original survey the surveyor probably does not have to deal with conflicting evidence because there are no original boundaries to be re-established. If the survey is a retracement survey, there will almost certainly be differences between the record location of monuments and their actual location on the ground. If the original survey is recent, properly performed and well monumented, discrepancies may be minor. If, however, the original survey is old, or monuments are missing, set in the wrong locations or moved, the calculations could prove to be difficult and time consuming. If many or all of the called-for monuments were not found, it may be necessary to do additional research and/or go back into the field until enough evidence is found to enable completion of the survey.

In a retracement survey, this phase of the survey is often considered the most challenging because, in difficult surveys, the surveyor must use experience and knowledge of the law of evidence, deed interpretation and the order of importance of title elements to properly calculate the true location of the boundaries. The surveyor will be required to use experience, judgment, and legal rules to decide what evidence should take precedence over other conflicting evidence.

It was stated at the beginning of this book, that although surveyors can measure miles within fractions of an inch, they are often unable to determine boundary locations within feet—or in some cases even tens or hundreds of feet. During the remainder of this chapter it will become apparent that the reason has little to do with measurement accuracy or ability. It has everything to do with understanding the laws governing boundary establishment. It should not be surprising that this aspect of a boundary survey can be far more difficult than any other phase. The following sections are primarily addressed to land in the eastern states and Texas, lands not in the Public Land Survey System. Lands in the Public Land Survey System have a specific set of rules for the establishment of lost corners. These rules are comprehensively described in the *Manual of Instructions for the Survey of the Public Lands of the United States*, as amended, so they need not be repeated here.

14.4 The Order of Importance of Evidence

A surveyor's duty is to locate a written description on the ground. But how should a surveyor proceed, when there are conflicting elements in the deed, when the physical evidence does not agree with the record information or when possession does not agree with the record location of the lines? Such problems are relatively common in boundary surveying and in retracement surveying in particular. The relative importance of title elements is well established by law. There are jurisdictional differences but they are generally as follows:

1. Unwritten Title
2. Senior Rights
3. Intentions of the parties to the conveyance
4. Calls for monuments
5. Distance
6. Direction
7. Area
8. Coordinates

Each of these will be considered in the following paragraphs.

How should a surveyor proceed, when there are conflicting elements in the deed, when the physical evidence does not agree with the record information or when possession does not agree with the record location of the lines?

14.5 Unwritten Title

Unwritten title to land is what is commonly referred to as title acquired by adverse possession. It is acquiring title to land belonging to another by occupying the land in a manner that excludes the rightful owner. The requirements for adverse possession are commonly established by statute and these will be discussed in another chapter. For our purposes here, suffice it to say that title to the land is transferred with no written instrument to evidence the transfer and that title so gained will take precedence over the written description.

14.6 Junior and Senior Rights

Junior and Senior Rights are rights which are acquired as a result of a sequence of conveyances made by a common grantor. The concept of junior and senior rights may best be illustrated by an example. Suppose Mr. Short owns a tract of land

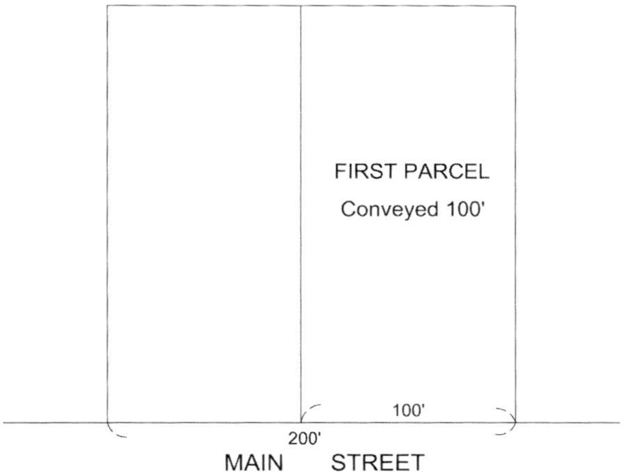

Fig. 14.3 Junior and senior rights

which he believes has 200 feet of frontage on Main Street. He decides to sell off two house lots; one this year and one next year. Each house lot will have 100 feet of frontage on Main Street. Mr. Short retains Mr. Cheepin Quick L.S., a land surveyor, to draw a plan to be used in creating the legal description of the first lot. Mr. Quick draws a plan showing 100 feet of frontage and Mr. Short uses this plan as the basis of a legal description to convey the first lot. The plan can be seen in Fig. 14.3. Right after the survey, the lot is sold to Mr. Primo. One year after the sale of the first lot, Mr. Short retains Mr. Quick again to draw a plan showing the remaining lot. Mr. Quick prepares a second plan, which shows 100 feet of frontage for the second lot. This second plan is used to create a deed description for the second lot, which Mr. Short then sells to Mrs. Sequent. Mr. Short has now conveyed two parcels of land, each having 100 feet of frontage on Main Street.

Junior and Senior Rights are rights which are acquired as a result of a sequence of conveyances made by a common grantor.

A few years after Mrs. Sequent acquires her lot she decides to put up a fence on the boundary between her lot and the lot owned by Mr. Primo. In order to save money, surveyor Cheepin Quik did not set any markers on the common boundary between Mr. Primo and Mrs. Sequent. Mrs. Sequent retains Mr. Slobut Thoreau, a surveyor, to monument the line so that she will know for certain she is constructing the fence on her own land. The deeds for the lots make no mention of monuments; they only say that each of the lots has 100 feet of frontage on Main Street. When Mr. Thoreau measures the frontage of both lots on the ground, he determines that there is only 190 feet of frontage, not 200 feet as called for in the record descriptions. How should Mr. Thoreau stake out Mrs. Sequent's lot? Should each lot lose 5 feet or should one of the owners lose the whole 10 feet?

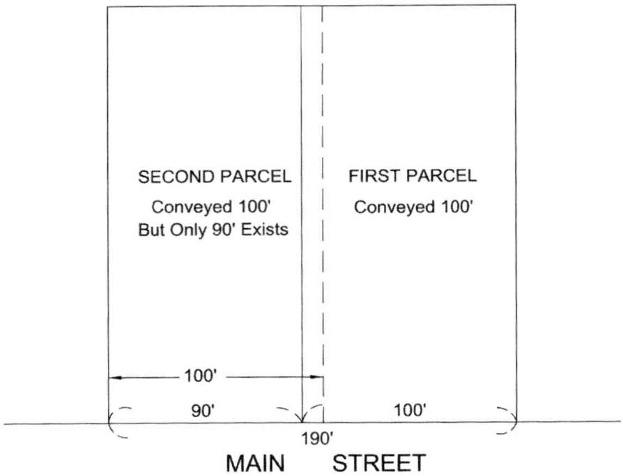

Fig. 14.4 Junior and senior rights

Mr. Short conveyed title to Mr. Primo before he conveyed to Mrs. Sequent. Mr. Primo's deed says that he owns 100 feet along Main Street. Because there was actually only 190 feet of frontage altogether, after the conveyance to Mr. Primo, Mr. Short was left with 90 feet of frontage, not 100 feet. It is well settled in law that you cannot convey that which you do not have. Mr. Short could therefore not have conveyed 100 feet to Mrs. Sequent because he only had 90 feet left to convey. The result is that Mr. Primo gets what his deed says he should have: 100 feet of frontage. Mrs. Sequent gets 90 feet. Another way to look at it is that, Mrs. Sequent gets whatever land is left. The results of Mr. Thoreau's survey can be seen in Figure 14.4.

In the above example, Mr. Primo is said to have **Senior Rights** because he acquired his land before Mrs. Sequent from a common grantor, Mr. Short. Mrs. Sequent is said to have **Junior Rights** because she acquired her land after Mr. Primo from the common grantor. Another way to state the relationship between Mr. Primo and Mrs. Sequent is that Mr. Primo's boundaries are determined by the metes and bounds description in the deed from Mr. Short. Mrs. Sequent's boundaries are determined by whatever land is left. If the surveyor, Mr. Thoreau, found that there was 210 feet of frontage altogether, not 190 feet, Mrs. Sequent would own the extra frontage—as long as the deed intended to convey all of Mr. Short's remaining land.

It should be noted that the metes and bounds description in Mr. Primo's deed only takes precedence over Mrs. Sequent's description where there is a conflict. If the other boundaries of both lots are not in conflict, the deed dimensions would be used. Another important point is that junior and senior rights only apply to land conveyed from a common grantor, or where the common grantor is in the chain of title. If Mrs. Sequent purchased her lot from Mr. Short's neighbor, who didn't acquire his property from Mr. Short or his predecessors in title, junior and senior rights would not apply. However, if Mr. Short died and his son inherited the land and it was the son who sold to Mrs. Sequent, junior and senior rights would apply.

14.7 Intentions of the Parties to the Conveyance

The phrase **Intentions of the Parties to the Conveyance** refers to what the grantor and grantee agreed to when the grantor sold and the grantee purchased the property. The intentions of the parties to the conveyance must be gathered exclusively from the deed itself. The writing takes precedence over what the parties have to say about it. Neither the grantor nor the grantee can contradict the writing. To allow a party to testify that some portion of the deed does not represent the understanding clearly expressed by the instrument is in effect a contradiction of the Statute of Frauds. This does not mean however, that parole evidence (spoken testimony) cannot be used to explain an ambiguity in a deed or explain circumstances in effect at the time the deed was made. Because the meaning of words can change over time, explaining what the words in the deed meant at the time of the conveyance is not contradicting the words themselves. It is merely placing the words in their original context.

Where elements of a deed contradict each other, it becomes necessary to ascertain the intentions of each of the conflicting elements. For example, suppose a deed states "541 feet to a stone bound marked 'WP'" and the stone bound is found at 520 feet not 541 feet. If the stone bound is clearly in its original location, the bound will control the location of the corner, and the distance will yield to that location. The reason is that monuments have priority over distances. It is probable that the monument more clearly expresses the intention of the parties than the distance. People can see monuments and they can see where monuments are located. Because people cannot see distances, the law considers distances to be more abstract than physical objects and therefore a less reliable indicator of the parties' intentions.

Where elements of a deed contradict each other, it becomes necessary to ascertain the intentions of each of the conflicting elements.

In the above example if the stone bound was now missing but there were people who knew the location of the bound, their testimony would be admissible as to the location of the bound and, if their testimony was believed, that location would prevail over the distance of 520 feet.

In general, deeds must be construed as a whole and no one part of a deed should take precedence over any other part.

14.8 Calls for Monuments

Monuments control over distances, direction and area. When an original survey is made, the surveyor has an ethical duty, and in some jurisdictions a legal duty, to set physical monuments locating the boundaries of the land being surveyed. The surveyor measures between the monuments and provides the owner with a metes and bounds description, usually in the form of a plan, showing the distances and directions of the boundary lines. This plan forms the basis for the description in the deed used to make the conveyance.

Monuments such as stone bounds are fixed in or upon the ground and are clearly visible by the parties to the conveyance (Fig. 14.5). Surveyors have been known to occasionally make mistakes in measuring, recording or writing

Fig. 14.5 Old stone bound made from a natural stone

directions and distances on plans. Lawyers sometimes make mistakes in interpreting plan dimensions and in writing bearings and distances in legal descriptions. As we learned in the last section, dimensions are more abstract than are monuments. Monuments such as stone bounds can be seen and touched. Dimensions must be imagined. For these reasons, monuments are considered more reliable than measurements and take precedence over them when there is a conflict.

Monuments set after a description is created are not controlling. The reason is that the parties to the transaction could not have seen monuments that were not yet set. An exception to this is where it can be proved that the monuments were set shortly after the conveyance at the exact location of the original boundary corner.

14.9 Dignity Among Monuments

Monuments called for in a legal description are generally to be weighed in the following order of priority.

1. Natural Monuments
2. Artificial Monuments
3. Record Monuments

The attributes of these various types of monuments have already been discussed. These priorities are based on the reliability and permanence of the monuments. The most reliable and permanent monuments take precedence over the less reliable and less permanent monuments.

14.10 Distance and Direction

In some jurisdictions distances are considered superior to direction. There are variations between the states and in some states direction takes precedence. In the sectionalized system of lands, primarily in the western U.S., distance prevails.

In construing a deed description having conflicting elements, the surveyor should hold as many terms in the deed as possible, and reject as few as possible. Since direction and distance are both needed to establish a boundary line, both must usually be considered. This rule is particularly true in metes and bounds descriptions which do not call for monuments. In such a case, bearings and distances are the only way to establish the boundary lines. In some descriptions, the surveyor will be able to mathematically determine the source of the error, thereby resolving the conflict.

14.11 Area

Area is the quantity of land contained within the boundaries of a parcel description. It is usually measured in square feet or acres. One acre contains 43,560 square feet. Where there is no metes and bounds description or where the metes and bounds description is ambiguous to the extent that area is the only potentially reliable dimension that expresses the grantor's intent, quantity prevails. For example, suppose a field of 5 acres was completely bounded by stone walls. The 5 acres was all the land that the grantor owned. The grantor conveyed the "southerly 3 acres" but provided no metes and bounds description. A surveyor wishing to establish the boundaries would measure along the stone walls until a line could be established running across the field that gave the south lot 3 acres and the north lot the remaining land. As the reader might imagine, in such a case the surveyor must exercise considerable discretion in locating the boundary and it is possible that another surveyor, perhaps working for the affected abutter, would locate the line differently.

As we have already discussed, before computers, area was sometimes difficult to calculate accurately. Because of this unreliability it carries a diminished weight in resolving conflicting deed elements. Where metes and bounds descriptions are ambiguous, and lacking a better method, area can sometimes be used as verification that the surveyor has correctly interpreted the deed.

14.12 Coordinates

Coordinates were traditionally considered the least reliable element because they resulted from calculations and not from direct observation. However, if a surveyor has calculated the coordinates of a monument and the monument later becomes lost or disturbed, the coordinates are evidence of the former location of the monument. If there is a conflict between the location of an undisturbed monument and its coordinates, the monument will prevail. The widespread use of GPS may be changing the importance of coordinates as a controlling element.

Many states have enacted statutes relating to state plane coordinates. These statutes provide that state plane coordinates are an acceptable means of locating a boundary corner and that the coordinates constitute a legal and satisfactory description of the corner. These statutes would seem to be supportive of an argument for the reliability of coordinates. Consider the following language from *The State Coordinate Systems (A Manual for Surveyors)*, Coast and Geodetic Survey:

99. *The description of a parcel of land for record purposes should have two aims: (a) It should positively identify the land for title purposes. (b) It should provide all necessary information for locating the land on the ground. These two aims are not identical: a description which satisfies the*

first may be wholly inadequate for the second. However, a single description to accomplish both aims can be prepared by adding to the description through which descent of title is traced, a technical description which includes the coordinates on the State system of one or more of the comers of the land.

102. *For a description of a public-land subdivision, an acceptable use of a State coordinate system would be the placing of the coordinates of the marked comers on an official plat, to which specific reference would be made in the written description. The plat should also, if practicable, show the coordinates of the control stations from which the coordinates of the subdivision comers were derived.*

103. *103. In other than public-land States, the inclusion of State coordinates in any form of description is also very desirable, except where there is an official plat on file showing the State coordinates, in which case a reference to the plat will make it part of the legal description. A carefully-prepared plat of an area embracing a number of land parcels and showing the State coordinates of the land comers and of control stations which define the State coordinate system is a very satisfactory description for survey purposes. A suitable reference to the plat in the description prepared for title purposes will make that description satisfactory for all purposes.*

We have seen that coordinate values derived from GPS observations are primarily dependent on the coordinates of the reference station(s) and not from the coordinates of other boundary or traverse points in the survey project. In other words, the accuracy of each point stands on its own. Using traditional linear traversing methods a location error of one of the points could propagate and affect the coordinates of other points. The location of a point from a properly established GPS location will have a known probability of error and this error is independent of errors in other points comprising the boundary corners of a property. Based on these considerations, it would seem that a boundary corner established using properly implemented GPS techniques should be just as reliable, perhaps even more reliable, than a boundary corner established using angles and distances.

As GPS is a new technology when compared to more traditional measurement techniques, it may take some time before the reliability of GPS measurements is recognized in the courts. Because the roots of real property law go back to feudal times, it has sometimes been said that property law is a constipated discipline. We now find ourselves besieged by rapidly changing technology, and this creates a great impetus for change. Like many other electronic technologies, GPS is continually improving and becoming more accurate and reliable. As the technology becomes more universally understood and accepted in the courts, we may eventually see the movement of coordinates from the last place in importance to being at least as important as distances and angles.

14.13 Example of Conflicting Deed Elements

Let us look at a simple example of a deed which has conflicting elements in the description. The example is based on the measurements of the garden fence, in Fig. 4.14, that we considered when we discussed property line closures. Recall that the fence dimensions were written down on a piece of paper and provided to a friend in a distant location so that they could construct a similar fence on their property. The deed description for the parcel reads as follows:

The grantor hereby conveys to the grantee a rectangular parcel of land, bounded and described as follows:

Beginning at a point in the northerly line of Old Pine Road, at Mile Marker 5;
Thence easterly in the northerly line of Old Pine Road seventy-eight (78) feet;
Thence northerly, by land of Ace Logging Corp., one hundred (100) feet;
Thence westerly, by last named land, eighty (80) feet;
Thence southerly, by last named land, one hundred (100) feet to the point of beginning;
Containing 8,000 square feet.

The first step is to make a sketch of the property from the deed description as shown in Fig. 14.6. The 90° angles are not specifically called for in the deed; however the deed does describe the parcel as being a rectangle, so the interior angles have to be 90°.

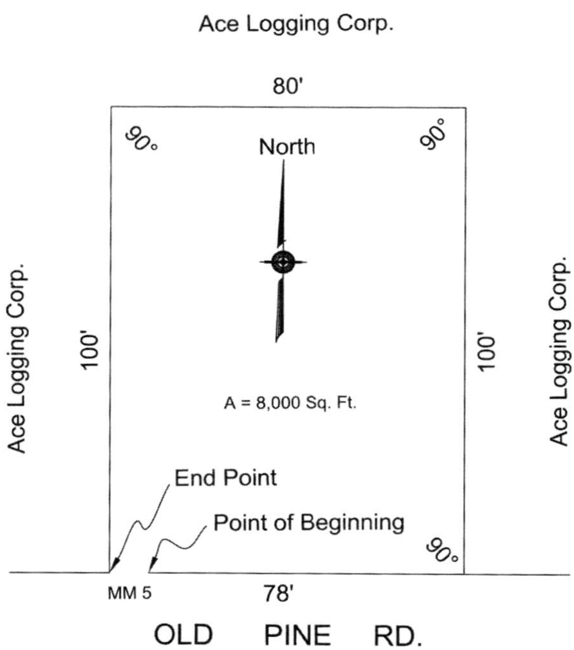

Fig. 14.6 Sketch of deed description with error

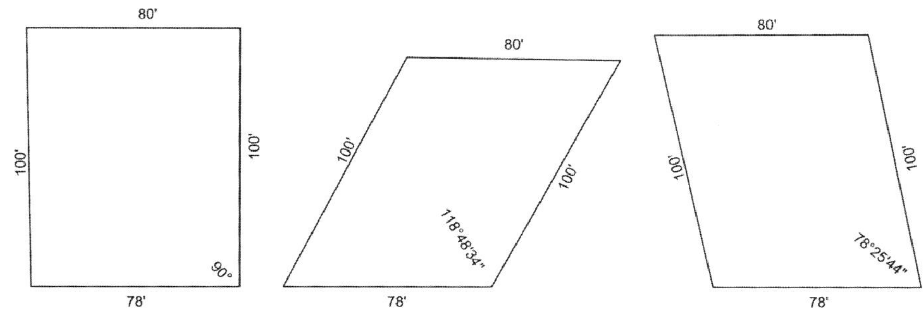

Fig. 14.7 Possible parcel shapes from holding dimensions

Notice the closure discrepancy at the southwest corner of the parcel. The boundaries of the property clearly do not close within acceptable limits. For example, if our minimum standard for an error of closure is 1 in 10,000, the largest error which would be acceptable in our parcel would be 0.036 feet. (Our perimeter is 358 feet so the allowable error of closure would be 358/10,000 = 0.036.) The error in this deed description is 2 feet so our actual error of closure is 358/2 = 179, or 1 foot in 179 feet. The actual error greatly exceeds the acceptable value. It is clear that the deed contains one or more errors, so our next step will be to consider some possible ways in which the deed might be interpreted, based on the rules we have just learned.

One option would be to discard either the 80 or 78 foot dimension and make both dimensions the same. We would then have a perfect closure with no error at all. However, based solely on the lengths of the boundary lines and the angles, both dimensions have identical probabilities of being wrong. There is nothing in the written description which would favor holding one of the lines over the other.

The rules governing the relative importance of title elements provide us with some guidance in how to proceed in this case. If we are in a state where distances prevail over angles, then we would have to allow distances to control the shape, not angles. So, one way of interpreting the deed description would be to hold the boundary lengths as fixed and ignore the angles. One potential problem with this approach is that, without angles, there is nothing to control the shape of the parcel. This can be seen in Fig. 14.7, where three possible configurations for the parcel have been shown. In fact, there are an infinite number of parcel shapes available, depending on the angle held at one of the corners. Holding a single angle will cause the shape to become a rigid figure. The problem with this approach is that we have no basis for choosing one angle over another.

We have also learned that where there are conflicting deed elements, we should discard as few elements as possible. Applying this logic and assuming we are in a state where distances control, we would hold the record distances and attempt to apply the 90° angle requirement. Because our shape is not a rectangle, it is only possible to hold the 90° angle in one of the parcel corners. The results are shown in Fig. 14.8. We have narrowed down the possible parcel configurations from an infinite number to 4

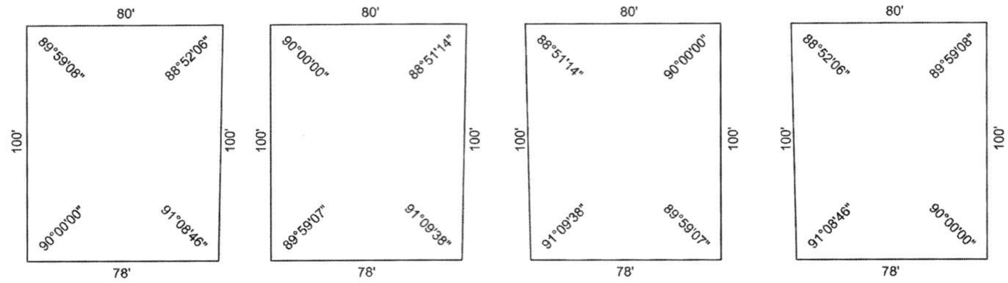

Fig. 14.8 Possible shapes holding angles

possibilities. The linear dimensions in the deed are held for all of the property lines. From left to right, each of the 4 parcels places the 90° dimension in one of the corners beginning with the southwest corner and ending at the southeast corner.

Although we have substantially narrowed our possibilities, we are still no closer to being able to locate the boundaries of the parcel on the ground because we have no reasonable basis for choosing one of the configurations over the other three.

Some might argue that the point of beginning should take precedence so that the final line should return to the point of beginning even if this means discarding the length and angle of the line. For example, in this case we would start at the point of beginning and use record angles and distances until we arrived at the northwest corner of the parcel. We would then just draw a line from that point to the point of beginning. There is case law to suggest that an error of closure should be placed in the last course. There is also case law which holds that the point of beginning should not have any greater importance than any other point. In the case at hand, we are not dealing with a typical error of closure which, as we have already discussed, would only amount to a few hundredths of a foot. Here we have an error of two feet, which seems much more like a blunder or omission than a measurement error. A blunder is a large error such as misreading a tape, misplacing a decimal point, transposing two numbers or some other gross, and sometimes obvious, error.

An additional interpretation would be to consider the deed elements as a whole in light of the intentions of the parties to the conveyance. Because the original description calls for a rectangle, it seems clear that the grantor intended to convey a rectangle. Unfortunately, this reasoning only puts us back where we started—requiring us to decide whether to hold 78 or 80 feet. However, this approach would at least have the benefit of allowing us to abandon the other theories which create skewed parcels. Clearly, if the grantor intended to convey a rectangle, we would not be honoring her intentions by creating some sort of skewed figure. We need to ask ourselves if there is any other information in the deed which might help to support our rectangle theory and if so, what dimensions we should hold.

Our deed description gives an area of 8,000 square feet. Recall that area is one of the last deed elements that should be considered as controlling. We already know that one of the reasons that area is last is because of its historical unreliability. Before computers were commonly available, calculating the area of irregularly shaped parcels was difficult and time consuming so areas tended to not be particularly reliable. As such, it was assumed that areas should usually not be controlling. An exception was where there was nothing better to rely upon. In the case before us, we presumably have a rectangle. Calculating the area of a rectangle would have been very simple, as it only involved multiplying the width times the height. So, in our case, the chances are very good that the area of this parcel, as described in the deed, is accurate. Here, the area is 8,000 square feet. This area supports an argument that the parcel should be a rectangle with sides measuring 80 feet by 100 feet. If 78 feet was the intended dimension, the area should have been shown as 7,800 square feet. In this case, it seems likely that the original surveyor would have had no trouble calculating the area and he would only have used 8,000 square feet if the parcel was, in fact, a rectangle. If we accept this premise, we would discard the 78 foot dimension and use 80 feet as the dimension along Old Pine Road. Our parcel would then have a perfect error of closure.

This relatively simple example is intended, in part, to demonstrate the uncertainty that surveyors have to deal with when construing deed descriptions with omissions and conflicting elements. Sometimes, there is a strong case for a particular line of reasoning. In other cases, the facts do not support such a high level of confidence and the conclusions are more tenuous. The conclusion in this case seems reasonable and it is probably correct. Keep in mind, however, that our result is still technically in conflict with the written description so another party, an abutter for example, may not agree and decide to challenge the interpretation, particularly if a different interpretation might favor their interests.

14.14 Apportionment in Subdivision Lots

A subdivision is the division of a parcel of land into lots. In zoning language, a lot usually means a parcel of land suitable for the construction of a dwelling house or other allowed building. Where a subdivision plan showing a series of lots, is recorded, the lots are considered to be created simultaneously. In contrast to sequential conveyances which give rise to junior and senior rights, the lots in a subdivision all have equal rights. The sequence in which the lots are conveyed has no effect on the rights of the lot owners in the subdivision.

Where a subdivision plan showing a series of lots, is recorded, the lots are considered to be created simultaneously. In contrast to sequential conveyances which give rise to junior and senior rights, the lots in a subdivision all have equal rights.

As an example of apportionment, consider Fig. 14.9. A subdivision plan was recorded showing a portion of a block bounded by three streets. Five lots were

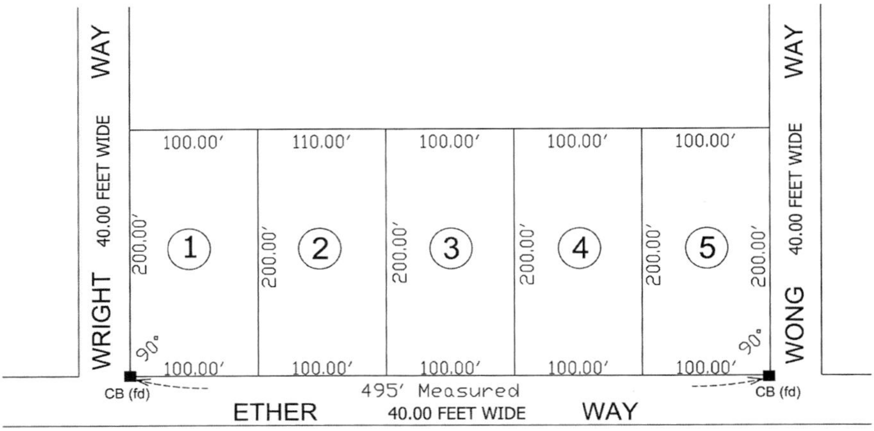

Fig. 14.9 Apportionment in a subdivision

created each having 100 feet of frontage on Ether Way. The plan shows that concrete bounds, noted CB (fd), were found at the street corners. The five lots were sold by the developer, one at a time over a period of two years.

A number of years after all five lots had been conveyed, the owner of lot 3 retains a surveyor to set the lot corners of her lot. The surveyor determines that the subdivision plan is called for in the client's deed and obtains the plan shown in Fig. 14.9 from the recorder's office. When the surveyor measures between the original concrete bounds he discovers that the distance is 495 feet, not the 500 feet called for on the plan. The monuments appear to be original, at their original locations and they are undisturbed. How is the deficiency of 5 feet applied?

Although the lots were sold one at a time over two years, the principle of junior and senior rights is not applicable in this case because the lots are considered to have been created simultaneously by the recording of the subdivision plan. It does not matter when the lots were sold.

In this case the surveyor will be required to apportion the deficiency among all of the lots. The deficiency is apportioned in proportion to the frontage of each lot. In this case, the apportionment mathematics is straightforward. The 5 foot deficiency apportioned over 5 lots each having 100 feet of frontage means that the frontage of each lot will be 1 foot less than its record dimension. Each lot will have 99 feet of frontage. To set the lot corners the surveyor would begin at either Wright Way or Wong Way and measure 198 feet (99 * 2) for the first corner of lot 3, then another 99 feet for the second lot corner. The remaining two lots would then have 99 feet each. It should be noted that apportionment is a rule of last resort. If the error is obvious, apportionment should not be used.

The procedure used above would also be applied if there were a surplus instead of a deficiency in the block. It should be noted that the surveyor held the 495 feet between the concrete bounds because these bounds were original monuments defining the boundaries of the street lines. When apportionment is used, streets are held in their monumented location according to the layout of the street. Street widths are not apportioned. In the above example the streets are 40 feet in width and the surveyor would not decrease the width of the street. Apportionment is limited to the block contained within the street boundaries.

The apportionment rule would not be applied where it can be shown mathematically that one or more of the dimensions on the plan was obviously in error. In the above example, the rear lot line of Lot 2 is shown as 110 feet not 100 feet. Angles and distances on the plan show that the lots are intended to be rectangles 100 feet wide by 200 feet deep. The dimension on the rear lot line of Lot 2 is clearly a typographical error. In this case the record dimension of 110 feet would be rejected and 100 feet would be used as the lot dimension (or 99 feet in the case of apportionment).

A minority of states have a rule called the **Remnant Rule** which places all of the error into an irregular end lot if one exists. In the above example if lot 5 were 105.64 feet in width while lots one through four were all 100 feet in width, the excess or deficiency would be applied only to Lot 5.

14.15 Boundary Agreements Between Abutters

We have looked at a number of factors which influence the location of boundaries. In many cases, reconciling the record and physical evidence will be very difficult and the surveyor will not be able to locate a boundary line on the ground with a high degree of confidence. In other cases the boundary location established by the surveyor will not be acceptable to one of the abutters. It was stated earlier that the only way to fix boundaries definitely is by a court order or by agreements between abutters.

Because of the high cost of litigation it is almost always preferable to attempt to enter into a boundary agreement with the abutter rather than to file a lawsuit. In some cases abutters will be unable to negotiate a boundary location acceptable to both parties and litigation is the only solution. Before entering into litigation the parties should realize that a court order is binding and the resolution may be one that will please neither party. This result will have been achieved only after tens of thousands of dollars in legal fees have been expended and several years have elapsed. In some cases stubbornness gets in the way of practical thinking and negotiation becomes difficult or impossible. In reality, most law suits that are filed end up being settled before trial. But even getting that far can still be very expensive. It is usually in the best interests of the parties to make a serious attempt to settle the issues before resorting to litigation.

If it appears that litigation is the only remedy, a party will need an attorney experienced in boundary dispute resolution. Boundary dispute resolution is a highly specialized area of the law. Although any attorney can take on the case, an attorney not experienced in property matters will need to spend considerable time getting up to speed on the substantive law which will certainly play a crucial role in the case. An attorney experienced in the discipline is more likely to be familiar with the applicable case law in the jurisdiction where the property sits.

If the parties are successful in reaching an agreement to locate the disputed boundary it will be necessary to create a written agreement signed by all of the parties having an interest in the land. It will then be necessary to place the agreement on record in the recorder's office. This agreement should be drafted by an attorney. Prior to drafting such an agreement the surveyor should set sufficient permanent markers on the lines which have been agreed to so that the abutters and future grantees will have no misunderstanding about the property line locations. After all, the present lack of permanent markers is probably why the parties are now being required to go through the trouble and expense of a boundary agreement. Consideration should also be given to using GPS (Global Positioning) to tie the boundaries into the state plane coordinate system as additional assurance that the boundary locations can be reproduced in the future. As already noted, some states have statutes which provide that use of state plane coordinates for the description of a boundary point is legally considered a full and adequate description of the location.

The surveyor should prepare a plan suitable for recording showing the line of agreement, the monuments set on the line and any other information which the surveyor believes would be helpful in reestablishing the line at some future time. This plan should be recorded along with the agreement. The agreement should call for the plan and give the book and page where the plan is recorded. The plan will thereby be incorporated into the agreement.

In some jurisdictions an agreement between abutters which is interpreted as agreeing to the true boundary may be overturned if it is later determined that the line established by the agreement is not in fact the true boundary. Where the parties agree to a line, mistakenly believing it to be the true line, the parties are not prevented from later claiming to the true line. To prevent such an occurrence it may be good practice to put words of conveyance in the agreement, whereby each party conveys to the other party any and all land lying on the other side of the line agreed to.

In some jurisdictions an agreement between abutters which is interpreted as agreeing to the true boundary may be overturned if it is later determined that the line established by the agreement is not in fact the true boundary.

If one or more of the parties has a mortgage on the land which will be the subject of the agreement, it will be necessary to contact the mortgage holder and obtain the mortgage holder's assent to the agreement. Mortgages often have "due on sale clauses" which make the note due and payable in full upon any transfer of the mortgaged property. If an agreement were recorded without the mortgage holder's consent and the agreement were later construed as a transfer, the land owner would probably be very upset to receive a demand for payment in full of the amount of the loan.

In most jurisdictions the rule of "first in time, first in right" is applicable so if a mortgage is recorded before the agreement, the agreement will be subject to the terms of the mortgage. If the mortgage is foreclosed the foreclosure proceedings may nullify the agreement unless the lending institution has assented to it.

An additional consideration when entering into boundary agreements is the possibility that an agreement may change the size of the lots which are being affected by the agreement. A change in a property line could make the lot area or frontage in violation of a city or town zoning ordinance. Property owners considering a boundary agreement should consult an attorney prior to entering into any verbal or written agreement.

14.16 Ownership of Streets and Ways

In some jurisdictions a conveyance of property abutting a street or way conveys a fee to the center of the way. A description having the language "bounded on the south by Main Street…" would convey the fee to the center of the way. However, language in the deed may exclude the fee in the way. If a deed read "thence running along the northerly line of Main Street…", the fee in the way would probably be excluded and no fee in the way will pass to the grantee. If the way is public, the fee is of little use to the abutting property owner as the public has a right of passage. If the way is private, the owner may be able to exclude certain people, but this will depend on the jurisdiction and the rights which others may have in the way. This rule is sometimes written into the state statutes and referred to as a "derelict fee statute".

Chapter 15
Setting Lot Corners

Once the boundary lines have been determined and calculated, the lot corners can be set. The surveyor makes a trip to the site, sets the instrument on a traverse point and uses the calculated angle and distance from a traverse line to the lot corner to set a marker at the exact location of the corner. If GPS is available, the surveyor just walks to the corner with the GPS receiver and fine tunes the location by gradually moving the receiver until the data collector shows it is exactly at the corner.

The actual monument that a surveyor will set depends upon where the lot corner falls. If the lot corner is in a lawn or open field or other location consisting of soil the surveyor can set a wooden stake, rebar, iron pipe or concrete or stone bound. If the lot corner falls on a large stone or stone wall, the surveyor is usually limited to setting a drill hole, copper or aluminum pin or a metal disk. A disk is often an aluminum disk about 2 inches in diameter that is held in place by a pin. It is identical to a metal pin except the large disk makes it easier to find (Fig. 15.1). A drill with a carbide bit is used to drill holes in stones and concrete for setting drill holes and pins (Fig. 15.2).

15.1 How Many Monuments?

It is good practice to set permanent monuments at each property corner and at 250–500 foot intervals along long property lines. It is not uncommon in retracement surveys to discover that many of the original monuments have disappeared or have been disturbed. The more original monuments that a retracement surveyor has available, the easier and more economical it will be to reestablish the original boundaries. In other words, the more monuments that a surveyor sets, the better the chance that a sufficient number will remain in the future to enable reestablishment of the boundaries. A property owner will incur substantial expense in the

P. Gay, *Practical Boundary Surveying*, DOI 10.1007/978-3-319-07158-9_15

Fig. 15.1 An aluminum disk held in place by a pin, making it easy to find

Fig. 15.2 A "John Henry" carbide drill and copper pins

performance of a survey. Reducing the number of permanent monuments often proves to be an unwise economic decision in the long run. In our age of short attention spans, multi-tasking and planned obsolescence, it is easy to lose sight of the permanence of real property.

Reducing the number of permanent monuments often proves to be an unwise economic decision in the long run.

15.2 What Kinds of Monuments should be Set

Permanent monuments should be of lasting materials, such as stone or concrete bounds or copper or aluminum pins and disks. Wooden stakes, even pressure treated ones, will not last. They usually do not go below the frost level, so they can be moved by freezing and thawing of the ground. Because of their small size they are easily moved by lawn mowers, garden tractors, etc. Although commonly used, in some surveyors' opinions small diameter iron pipes and rebars are also poor choices for permanent boundary corners, particularly if they are not sunk deeply into the ground. An exception might be a large iron pipe 3–4 inches in diameter set in concrete several feet deep (Fig. 15.3). Because of their small diameter and uniform shape, pipes and rebars are relatively easy to pull out of the ground. Where boundary disputes exist, it is sometimes discovered that these markers have made nocturnal journeys to locations more favorable to one of the disputants.

Some jurisdictions require monuments to have a magnetic signature so that they can be readily found using a metal detector. This usually requires that the monument have a quantity of ferrous material such as iron or steel. Such a requirement would not permit the use of stone bounds (unless they were composed of rock containing a magnetic mineral such as magnetite). Iron and steel, particularly in wet climates, is subject to corrosion so unless the marker is treated by galvanizing or other protective means, the long term viability of such markers may be open to question. Some statutes include language wisely allowing the surveyor discretion in selecting the most appropriate and durable monument under the circumstances. It could be argued that it is the surveyor who is in the best position to know the ground conditions of a particular boundary corner and the appropriateness of a particular marker for those conditions. It may also be possible to add permanent magnets to otherwise nonmagnetic markers in order to comply with a statute, or metal pipes or rebars as witness markers.

When setting lot corners with a total station, the instrument will usually be set up on a traverse point somewhere near the point to be set. The instrument person will zero the instrument on the backsight and turn an angle. This will be the line on which the point is located. The rod person will then be placed on line by the instrument person. Often a prism pole is held on line so that a distance can

Fig. 15.3 A pipe filled with concrete, probably set by homeowner

be taken. The difference between the shot and the distance to the lot corner will tell the rod person how far to move, either toward or away from the instrument in order to be at the lot corner. The rod person usually has a short tape measure that can be used to measure the distance from the shot to the lot corner once he is within a couple of feet of the corner. If a concrete bound is to be set at the lot corner, a wooden stake is often driven in the ground first. Once the stake is in the ground, the quickest and most certain way to place a point on the stake is to use a plumb bob to put a point exactly on line near the front of the stake as shown in Fig. 15.4, "First Mark". Another point is placed near the rear of the stake and a line drawn between them. Once a distance shot is taken on one of the points it is a simple matter to use a tape to measure the distance along the line to the lot corner. Of course this assumes that the rod person hammered the stake into the ground in the correct location so that the final point falls on top of the stake. This seems to be such a simple procedure that it is a wonder that more people do not use it instead of taking a half dozen shots or more with the prism pole—guessing at each shot - before the final point is eventually located. The same procedure can be used for setting points on sidewalks, stones or nearly any surface.

It was just noted that if a concrete or stone bound is to be set, it is customary to set a point in the ground first. The point is commonly a wooden stake. The location of the stake tells the person digging the hole for the bound exactly where to dig. Before commencing to dig, two stakes are set as ties to the corner as shown in Fig. 15.5. The stakes should be about 90° apart and 3–6 feet out from the lot

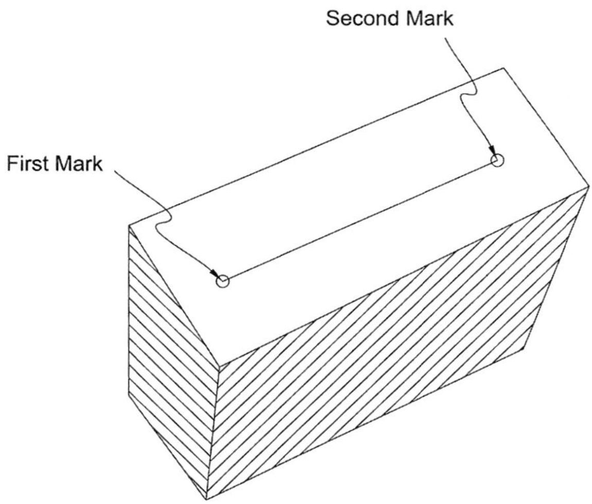

Fig. 15.4 Using a line to set point

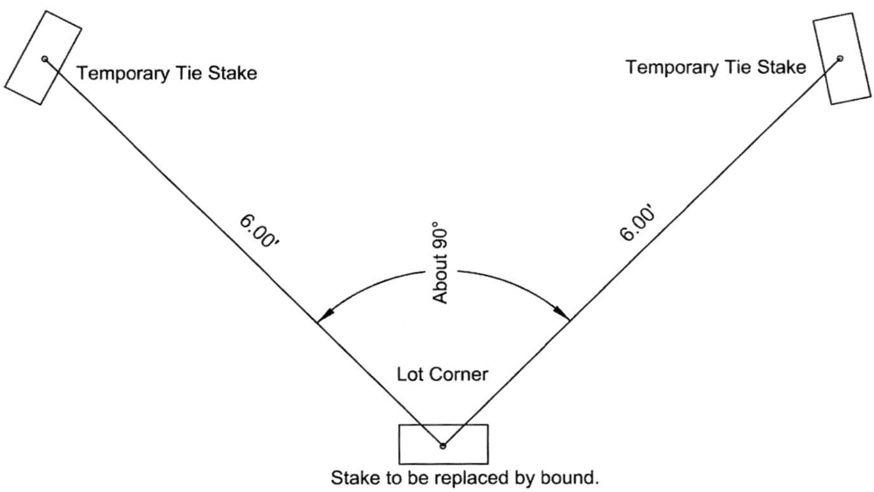

Fig. 15.5 Ties to a bound

corner. If the soil is likely to contain stones, longer ties are best because the hole may need to be enlarged in order to remove the stones. It is easiest to use the same distance for each tie. Using an even foot makes it easier to remember the distance, so it needn't be written down. Once the bound has been set, the tape can be used to measure out from the ties to the top of the bound for setting the drill hole location. The distance is measured from each tie. The original point will be where the two distances intersect. Setting a concrete bound can be seen in Fig. 15.6.

Fig. 15.6 Setting a concrete bound in a dry hole

Setting bounds in loose sandy soil such as on a beach can be quite a lot of work because the sand from the sides of the excavation causes the sides to cave-in as the hole becomes deeper. In such cases, the top of the hole can become quite large and a substantial amount of material ends up being removed before the hole achieves the necessary depth. Digging a hole in wet areas can also be a problem because of the water running into the hole turning the dirt into mud.

The author recalls his first experience setting a bound as a teenager fresh out of high school working as a rodman for a small rural surveying firm during the summer. Both the party chief and the transit man had many years of experience surveying. I was just beginning to learn some of the basic duties required of a rod man. The party chief and transit man said they would dig the hole if I would walk back to the truck, throw a concrete bound on my shoulder, and bring it over to the hole. The day was hot, the distance considerable and the bound very heavy but I finally staggered back to where the crew was standing. I was told that the hole was ready and to just drop the bound in. Out of breath and with sweat burning my eyes and impairing my vision, I failed to perceive that the party chief and transit man were standing quite far from the hole. I also didn't notice the eager anticipation on their faces as I approached. By that time, the weight of the bound on my shoulder was causing considerable discomfort and I was anxious to be rid of it so I didn't hesitate to follow their instructions and drop the bound into the hole. The party chief had presumably forgotten to tell me that the bottom of the hole was full of water. I don't know what the speed of the bound was when it hit the water but, like the large meteorite which smashed into the Great Dismal Swamp, creating Lake Drummond, it must have reached terminal velocity. I was instantly and completely covered in mud. The success of their well-planned and skillfully executed experiment clearly delighted the party chief and transit man. For the rest of the summer, they took great pleasure in giving an animated description of the events to anyone who was willing to listen.

In some cases it is possible to set many of the lot corners during the traverse portion of the survey. An example would be where some of the property lines are stone walls. Where a deed calls for a wall as a boundary, the boundary line is usually the middle of the wall. As noted earlier, an exception to this rule, in some jurisdictions is where the wall is a boundary between private property and a road or cemetery. In such cases the street face of the wall or the outside face of the cemetery wall is the boundary.

Stone walls, particularly old farm walls, are often far from straight. It is the surveyor's duty to take enough locations of the wall so that the center line or face, as the case may be, can be described accurately. If the wall is very crooked this may require taking locations of the wall every few feet. A surveyor will often set drill holes or pins at many of the locations where the wall makes a jog or bend during the traverse phase of the survey. Because the instrument is already set up on a nearby traverse point this is often the most efficient way to set the monuments. Many old walls are built from small field stones which are easily disturbed. In such cases, when a surveyor comes across a large base stone, it should be used to set a drill hole or pin. If the monuments are not set during the traverse phase but calculated in the office, then set on a return trip to the field, the surveyor will have no choice about where to set the monument. If it falls on a small stone it will be necessary to rebuild that section of the wall with large stones—a time consuming and expensive process. In such cases it is much easier to set markers in a wall during the initial traverse phase of the survey.

Stone walls, particularly old farm walls, are often far from straight. It is the surveyor's duty to take enough locations of the wall so that the center line or face, as the case may be, can be described accurately.

Fig. 15.7 Using GPS at a lot corner

In many instances, the location of a boundary corner is not suitable for a permanent marker. As an example, the preceding paragraph described a stone wall made from small, easily moved stones. In such cases it is often a greater service to the surveyor's client to set a permanent marker in a large immovable stone in the vicinity of the corner. The survey plan can then show the relation between the corner and marker with a bearing and distance. If the permanent marker is relatively near the lot corner, the corner can be easily replaced at any time in the future. This method is arguably much better than relying solely on an impermanent marker set at the corner which is very likely to become disturbed or lost.

15.3 Using Global Positioning for Setting Corners

The widespread use of GPS has the potential to make reestablishment of lot corners less costly than traditional methods. Once a survey has been tied into a state plane coordinate system, replacing a missing lot corner can be much less time consuming than when using traditional techniques, if a long traverse must be run to control monuments (Fig. 15.7). Using RTK GPS, the rover only requires an occupation time of a few seconds or minutes. Because of the accuracy limitations discussed earlier, RTK may not be suitable for boundary establishment if the required level of accuracy is not attainable. It must be kept in mind that the receivers are still subject to accuracy limitations caused by objects between the antenna and the satellites so it may still be necessary to use conventional surveying techniques to set some of the lot corners (Fig. 15.7).

Having the boundaries related to fixed datum certainly makes sense, particularly for large parcels of land. For small lot surveys, the cost may still be prohibitive, but that is rapidly changing as technology improves and equipment cost decreases. Although GPS may make it easier to replace missing lot corners it will still be desirable to set permanent markers at as many lot corners as possible because doing so will almost certainly be more cost effective than paying a surveyor every few years to replace missing or disturbed corner markers. In summary, while tying a survey into a state plane coordinate system is desirable, it does not take the place of on-site permanent markers.

Chapter 16
Survey Plans

A **Survey Plan** shows the results of a boundary survey. Survey plans are also called **Plats** and **Maps**. Survey plans are drawn in two dimensions, on a horizontal plane similar to an aerial photograph taken from directly overhead.

Not all boundary surveys result in the creation of a plan. If the purpose of the survey is to simply set the corners of an existing residential lot (a retracement survey) and there is already a plan on record, there would probably be no reason to draw another plan, unless the jurisdiction requires one. If the retracement survey discovered issues with the original survey or original plan, then it might make sense to draw a plan showing the survey results. Keep in mind that a retracement survey cannot change the original boundaries, but if mistakes were made in the original plan or the survey discovered ambiguities in the original survey, a new plan can be used to memorialize these facts.

Keep in mind that a retracement survey cannot change the original boundaries, but if mistakes were made in the original plan or the survey discovered ambiguities in the original survey, a new plan can be used to memorialize these facts.

If the purpose of the survey is to create new boundaries, then a plan will need to be drawn showing the new lines. The plan will be used as a basis for creating a deed for the new parcel or parcels if the survey created a multi-lot subdivision. In this case, depending on the jurisdiction, the plan may have to go before a municipal planning agency before the plan will be accepted for recording.

Sometimes plans are required by permitting departments to show the proposed locations of buildings and other improvements. A plan may be required

P. Gay, *Practical Boundary Surveying*, DOI 10.1007/978-3-319-07158-9_16

by a conservation agency for wetlands delineation. A plan may be required by a bank before it will approve a mortgage or other financing. In some circumstances ALTA/ASCM (American Land Title Association/American Congress on Surveying and Mapping) requirements must be met, which include surveying standards and plan standards. In our increasingly bureaucratic world, municipal, state and federal agencies appear to have a limitless appetite for enacting regulations which require the applicant to submit plans for just about any imaginable project or purpose.

As the preceding paragraphs indicate, depending on the circumstances, a plan may be regulated by a government authority. So, the plan purpose and jurisdiction may be the overriding factor in determining whether or not a plan will need to be drawn, and if so, what information will need to be shown.

16.1 Recording Plans

Plans can be prepared for recording in a public recorder's office or merely held in private records.

The recording of plans is yet another area that will depend to a certain extent on government regulation and therefore the jurisdiction. As has already discussed, the land records recording system primarily serves a notice function. When a deed or plan is placed on record, the public is presumed to be on notice of the contents of the document. The recording office usually does not take any position relative to the correctness of the information shown on the plan. They simply accept documents for recording (along with a recording fee) from any person who wants the document to be recorded. That is not to say that all documents will be accepted however. With regard to plans, recorders offices often have regulations such as plan size, plan material, ink type, whether the plan is stamped and signed by a licensed land surveyor, whether it needs to be signed by a planning board and other items not directly related to the survey. The recorder's office takes no position relative to whether the survey information shown on the plan such as the metes and bounds, monuments, abutter names and other information relating particularly to the surveyed parcel, is correct or incorrect.

16.2 Effect of Recording a Plan

Let's consider the effect of placing a plan on record. Imagine that a surveyor performed a retracement survey of a 100 year old parcel described ambiguously by the client's deed. The abutter's deed is no more helpful in describing the location of the common boundary. In spite of these difficulties, the surveyor is confident that her opinion regarding the location of the boundary is reasonable and will stand up to scrutiny. The abutter disagrees with the surveyor's boundary location but presently doesn't want to go to the expense of hiring "his own surveyor". The client has confidence in the surveyor's results (after all he did pay him quite a lot of money to perform the survey) and wants to place a plan on record showing the results of the new survey. Will placing a plan on record fix the disputed boundary and put the matter to rest? Probably not.

Recall that surveyors do not have the authority to establish boundaries with the force of law. They can only give their opinion as to where they think the true boundary is located. In this case, it seems there is a degree of uncertainty about the real boundary location and another surveyor may view the evidence differently and come up with a different interpretation of the location. Merely placing a plan on record will not change this.

One possible advantage to recording a plan in this case might be that it would serve as public notice to the abutter that the land owner is claiming to a certain line. If the plan showed the conditions on the ground, including the monumented line, the land owner could, after the statutory period, claim the boundary location was settled through adverse possession or acquiescence. Recording the plan may be useful to the land owner for the purposes of establishing the conditions on the ground at the time and the date on which her possession began. The abutter would have difficulty claiming a later date because he would be presumed to be on notice of the recorded plan. Of course, the owner would still need to prove the remaining elements of her claim. Also, the plan, by itself, would have no bearing on the actual possession because the possession has to be physically on the ground.

16.3 Plan Materials and Size

Many plans, in years past, were drawn on paper of some sort, in either pen or pencil. Modern copiers are capable of creating an image from an opaque sheet. Prior to this technology it was necessary to draw plans on a translucent or semi-transparent material so that light could pass through for the copy process. Tracing paper was often used as it was relatively inexpensive. However, tracing paper is not particularly durable.

Plans which needed to be durable would often be drawn on **Drafting Linen**, sometimes called **Drafting Cloth**. The linen had a very fine weave and was starched in order to provide a smooth drafting surface. Drafting linen was strong and durable. When properly prepared it would accept ink. It could also be erased. It was not waterproof. If a piece of it was laundered to remove the starch, it would produce a wonderfully fine fabric.

Today, most plans intended for long life and durability are drawn or plotted on **Drafting Film**, usually made from a polyester material. Polyester is very strong and stable. It will not tear easily and it is not affected by moisture. As it is translucent, it can be copied using older technology copiers.

The size of the plan will be dependent on the amount of information to be shown. Fonts should be large enough so that words and numbers are clearly visible. A minimum text height about 0.1 inches is a good place to start. Keep in mind that plans may be reduced in size for convenience so very small fonts may become illegible. If the plan is to be submitted to some agency or recorder's office there may be maximum and minimum plan sizes. There may also be minimum letter sizes. Because most plans are now created in CAD programs, adjusting the size is quick and easy.

16.4 Information Shown on a Boundary Survey Plan

The information which is shown on a plan will be dependent on the plan's purpose. If the plan is to be submitted to a planning board or other agency, they will probably have their own specific requirements. Even if an approving agency or statue requires certain information to be shown on a plan, there are some general rules and good practices to be considered when drafting a plan. This section is not intended to be an exhaustive treatment of drafting survey plans, it is only intended to provide the reader with a general introduction to some of the more important points.

16.4.1 General Requirements

North should point generally toward the top of the sheet as in Figs. 16.1 and 16.2. A north arrow must be shown on every drawing and the arrow should be labeled showing the basis of bearing system used as in Fig. 16.3. In this plan the north arrow is labeled *Magnetic*.

The plan must be drawn to scale and the scale noted on the plan. In most CAD programs the units are real world units so everything is automatically to scale. Scale notation is often done in the title block as in Figs. 16.1 and 16.2. If it is necessary to show a detail of some area, the detail should show the scale or, if the detail is not drawn to scale, it should be labeled "No Scale". Because a plan can be reduced or enlarged using a copier it is a good idea to show a graphic scale. Fig. 16.4 shows a typical graphic scale. When a graphic scale is drawn on the plan, shrinking or enlarging the plan also shrinks or enlarges the graphic scale to it can always be used with a divider or ruler as an aid to estimating distances on the plan.

If the location of the parcel is not apparent from the plan itself, it is usually a good idea to add a **Locus Map**. A locus map is a small map on the drawing showing the distance from the nearest road intersection or some other prominent feature which would help someone unfamiliar with the locus to find it.

Abutter's names should be shown on the plan as in Fig. 16.3. Sometimes the book and page of the abutter's deed is also shown. It is customary to show dashed lines dividing the ownership of abutting properties, as in Fig. 16.3. This differentiates them from the locus boundary lines. Sometimes this would be obvious but other times it is not. Abutting properties should usually not be dimensioned, unless there is a clear reason for doing so. The surveyor was not retained by the abutter so liability may be incurred. If a monument on an abutting property was used, the surveyor should be careful to indicate that the dimensions from the locus to the monument are labeled as tie lines so they will not be interpreted as boundary lines.

Streets should be labeled as **Public** or **Private**. The street width should also be shown. If the width of the street varies or is unknown, the street can be labeled **Variable Width**. Figure 16.5 shows a convenient way to do this. If there is a

Fig. 16.1 Survey plan of a small parcel of land bounded by a wall

Fig. 16.2 Plan showing a retracement survey

public layout of the street, this should be noted along with the layout year and whether the layout is a town, county or state layout. All monuments used for fixing street lines should be shown. The right side of Fig. 16.5 shows two monuments at an angle in the street. Another monument is shown on the left side of the drawing marking a point where the street begins to curve.

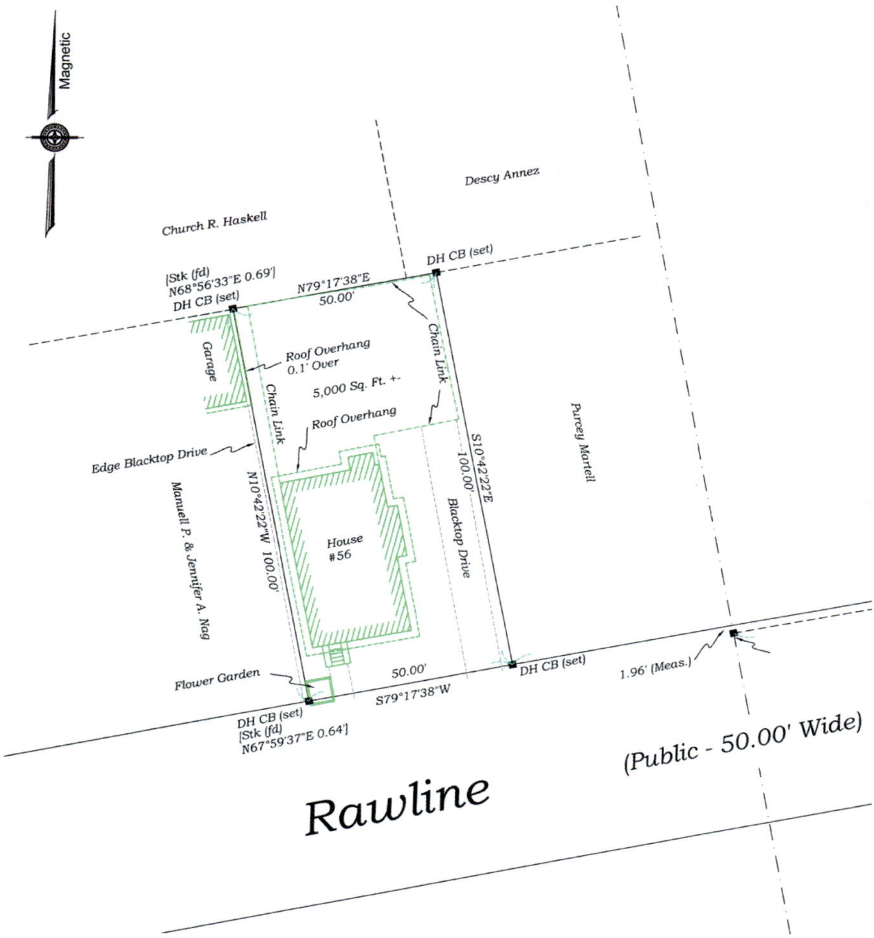

Fig. 16.3 Portion of a retracement survey plan

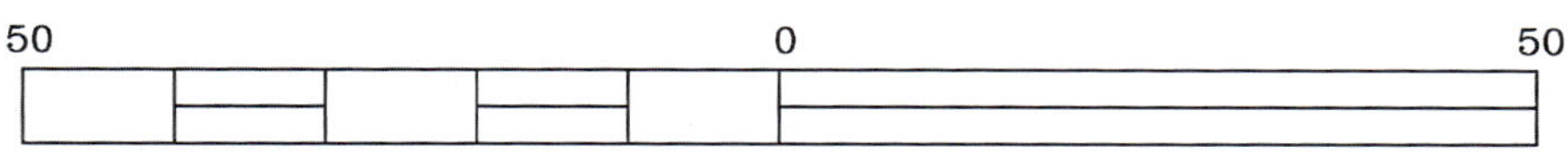

Fig. 16.4 Graphic scale

Ways over the locus or other signs of possible adverse use should be shown. A surveyor may incur liability by not bringing evidence of such uses to the attention of the client or someone who might rely on the plan. One way of doing this is to show the extent of occupation on the plan.

Fences, walls or other features which appear to be physical evidence of ownership should be shown along with distances from the feature to the boundary lines, particularly when such features are close to boundaries.

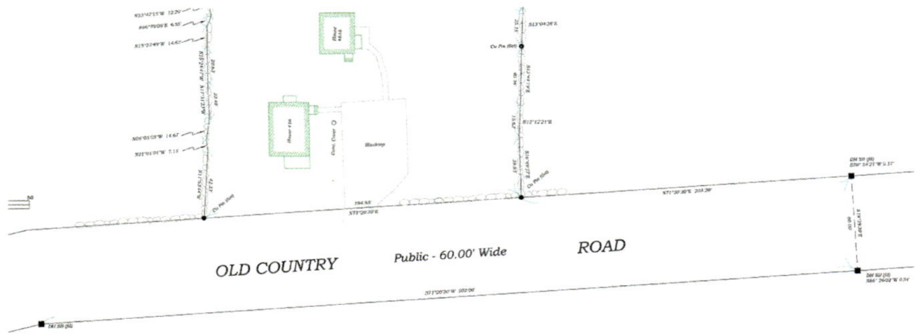

Fig. 16.5 Street name, width and designation as a public way

Record easements should be shown if they are over the locus or appurtenant to the locus.

Buildings and material improvements on the locus should be located and shown on the plan. Buildings on abutting property which encroach on the locus must be shown. Buildings not on the locus, but close to the locus boundaries, should also be shown, particularly if the location of the boundary is uncertain or disputed. Some surveyors show the portions of buildings within 10 feet of locus boundaries. Surveyors should be on the lookout for roof overhangs on abutting property which might encroach on the locus even if building walls do not. The roof of the garage at the northwest corner of the lot in Fig. 16.3 overhangs the boundary by 0.1 feet. Notice that the garage walls are clear of the boundary.

Water courses should be shown and labeled if they have a name. Direction of flow should be shown. If known, indicate whether or not the flow is seasonal.

16.4.2 Boundaries

Obviously, the boundaries of the subject property must be shown on the plan. Boundaries of the locus should be drawn in such a way as to clearly indicate that they are the lines of ownership. If the boundary lines could be confused with other lines, such as the boundary lines of an abutting parcel, there should be some distinguishing feature to the lines. For example, they could be drawn using a heavier line weight or the interior of the locus could be shaded. Some surveyors like to draw the lines of a street or way using a heavier line weight to differentiate it as a street line. In many cases the abutting property owner owns to the center of the way but this is rarely shown on survey plans.

Each boundary should be labeled with a bearing and distance. Where boundaries are not straight, each segment must have its own bearing and distance. Tick marks should be used to indicate which line segment the bearing and distance applies to. It is customary to place the distance inside the locus and the bearing outside the locus, but doing it the other way around is not fatal. If the boundary contains many short segments, it may be necessary to show the bearing and distance away from the line and use an arrow or leader line pointing to the boundary segment. See Fig. 16.6 for examples of a series bearings and distances along meandering wall. Where space permits, bearings and distances are written along the line segment. When there is not enough space, a leader line from the dimension to the line is used to clearly indicate which line segment is being described.

Where possible, bearings should flow continuously around the parcel, either clockwise or counterclockwise. This is important because a deed description will probably be created from the plan. A deed description has a point of beginning and, from that point, each bearing and distance is described in sequence all around the parcel back to the point of beginning. Obviously, having a line with a reversed bearing would incorrectly describe the parcel.

When the plan shows a subdivision with parcels having common boundaries, it will not be possible to maintain this continuity unless double bearings are used on the common lines. This not commonly done however and may lead to more

Fig. 16.6 Bearings and distances along a boundary wall

confusion than having a single bearing on the line. In this case it will be up to the attorney or person drafting the deed to correctly modify the quadrant of the bearing.

Where boundaries are along water courses, one or more tie lines should be used so that a future surveyor will be able to verify the boundary closure. Tie lines have already been discussed and an example is shown in Fig. 13.3.

The area of the locus should be shown. Notice in Fig. 16.3 that the lot area is shown as 5,000 Sq. Ft. ±. It is customary to show area as plus or minus. This practice probably started when it was difficult and time consuming to precisely calculate areas of irregular parcels. Now, computers are routinely able to calculate precise areas. Another possible reason is that errors in surveying may result in a client either losing or gaining property with potential legal claims against the surveyor. Property is valued in part by its area. Other things being equal, a larger parcel is worth more than a smaller one. Monetary damages for a surplus or deficiency may be calculated by the change in area. By showing area as "more or less", it might give the surveyor some wiggle room in a suit. The result might be to mitigate damages being assessed against the surveyor. That said, there are probably safer things to bet on.

The plan should be signed and sealed by the licensed surveyor.

16.4.3 Monuments and Control

The plan should show any survey control that was used to establish the boundaries or used to tie the survey into. This would include monuments, either of record or not of record, street monuments and, if applicable, ties to the state plane coordinate system and vertical datum used to establish a boundary along a water course or shoreline. Traverse lines are not usually shown on a survey plan unless they are required to be shown by a government agency or if the plan is a work plan for use by another surveyor. For example, in Massachusetts, plans submitted to the Land Court must show traverse lines colored in red.

The plan should show any survey control that was used to establish the boundaries or used to tie the survey into. This would include monuments, either of record or not of record, street monuments and, if applicable, ties to the state plane coordinate system and vertical datum used to establish a boundary along a water course or shoreline.

The importance of setting permanent monuments on boundary corners and lines has already been discussed. Original monuments which were found must be noted. If an original monument is called-for in the record but is now missing, this should also be noted. Monuments should be described on the plan with the following information:

Monument type: Concrete Bound (CB), Stone Bound (SB), Pin.
Type of point in monument: Drill Hole in CB, Copper Pin in SB.
Whether the monument was Found, Set or Not Found.
Some examples are:
SB (fd.)—Stone bound found. There was no drill hole or other mark in the bound.
DH CB (fd.)—Drill hole found in concrete bound.
DH SB (set.)—Drill hole set in stone bound. This usually means that both the bound and drill hole were set.
DH set in SB (fd.)—This would indicate that the stone bound was found and the drill hole was set in the bound.
Cu Pin (set)—Copper pin set.
Al Pin (set)—Aluminum pin set.
DH (fd.)—Drill hold found.
SB (fd.) B&C—The back and center of a stone bound found.
If the precise location of an existing monument was held it should be so noted as follows:
DH CB (fd.) Held.
If a point off the monument is held, for example if a monument has been disturbed, it should be so noted along with a bearing and distance from the point held to the monument. See Fig. 16.7 for example:
DH SB (fd) S86° 26' 02"E 0.54'. This means "A drill hole in a stone bound was found South 86 degrees, 26 minutes and 2 seconds East, at a distance of 0.54 feet from the corner".
Any other information that is specific to the monument. For example "Town line Bound".

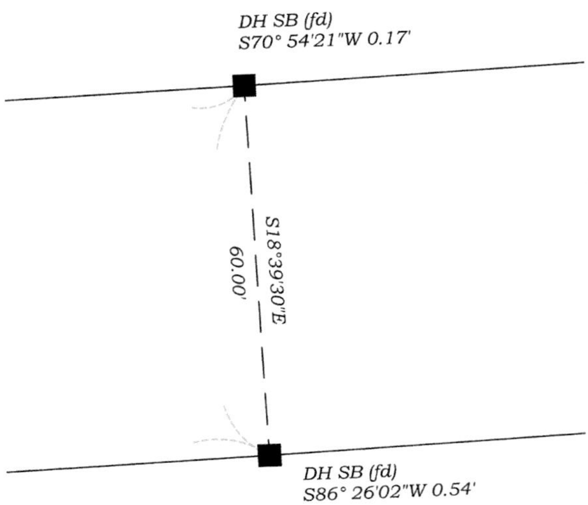

Fig. 16.7 Monuments found offset from a street angle point. Note bearing and distance showing distance from angle point to monument

Sometimes a monument is called-for in the record but has been destroyed. If the surveyor makes an exhaustive search for the monument and determines that it is gone, it should be shown on the plan but labeled "not found" or "record stone bound not found" or some other language to that effect. The label serves as notice that the surveyor did look for the monument but was unable to recover it. This indicates that the surveyor exercised due diligence in searching for the monument.

If the surveyor makes an exhaustive search for the monument and determines that it is gone, it should be shown on the plan but labeled "not found" or "record stone bound not found" or some other language to that effect.

The fact that many permanent monuments end up missing a few years after they are set, is a good reason to consider using additional means of fixing lot corners. One easy but often overlooked method is to tie boundary corners into buildings on the locus. Many total stations have reflectorless distance measuring built in. This makes it possible to locate building corners in a second or two with a high degree of accuracy. Although some surveyors do not like to show coordinates on a survey plan, it is an easy and clear way to connect a survey to buildings, particularly foundations which often have sharp corners and are not likely to be affected by re-shingling or new wooden corner boards on a house. Alternately, a bearing and distance can be shown from the building to the lot corner. However, tie lines like this can quickly clutter a plan, particularly if there are many of them.

Chapter 17
Unwritten Title

It is possible to own land even though there is no writing to evidence it. This is called **Unwritten Title.** Unwritten title is an exception to the Statute of Frauds which, as we have learned, requires that all conveyances of interests in real property be in writing. Unwritten title takes precedence over written title. It can come into being through adverse possession, prescription or by virtue of verbal boundary agreements.

Although unwritten title takes precedence over written title, the claimant must go to court to prove the title before it can be adjudicated. A lawsuit costs money, takes time, and even where the claim has substantial merit a favorable outcome is far from certain. So, from a practical point of view, even if the claimant has acquired the property interest it may be of little economic value until a court has issued an order or decree affirming the claimant's rights.

17.1 Adverse Possession

Adverse Possession is a species of unwritten title. It is taking property belonging to another without their consent. The adverse possessor acquires title by occupying the land of another person. In order to prevail the occupation must be wrongful. Each state has their own statutory or common law rules establishing the requirements for adverse possession, but in most jurisdictions, the possession must meet the following conditions. The possession must be:

1. Actual.
2. Open and notorious.
3. Hostile
4. Exclusive
5. Continuous

P. Gay, *Practical Boundary Surveying*, DOI 10.1007/978-3-319-07158-9_17

Adverse Possession is a species of unwritten title. It is taking property belonging to another without their consent.

Possession must be **Actual**. There must be some sort of physical occupation of the property. Building a fence on an abutting property would be actual possession. So would erecting a building. Mowing a lawn, planting shrubs or paving a driveway could be enough to constitute actual possession. Merely putting a plan on record, without more, does not constitute actual possession because there is no physical occupation of the land.

Possession must be **Open and Notorious**. Possession has to be visible. In addition to requiring physical occupation, the occupation must be open and notorious. The possession must be such that the rightful owner of the property would be on notice of the possession. The means of occupying the land must be visible in such a way that the owner would see the possession were she to inspect the property. The owner needn't actually see the possession however. The fact that the owner lived in another state and failed to inspect the property would not exempt the owner from the requirement. The possession may be deemed open and notorious if it would be obvious to any person inspecting the property. Possession would not be open and notorious, for example, if the possessor were to erect marks of possession only at night and remove them during the day.

The possession must be **Hostile** or adverse to the interests of the owner. This generally means nothing more than saying that the possession is without permission of the rightful owner. Permissive use defeats adverse possession. If a property owner allowed a neighbor to erect a fence on the owner's side of the common boundary by giving the neighbor permission to do so, the neighbor could not thereafter claim adverse possession of the land between the fence and the record boundary.

In a few jurisdictions the state of mind of the possessor is material in determining hostility. If the possessor believes that he is occupying to the true line then there can be no hostility and therefore no adverse possession. The majority of states view this as a bad rule. It should be enough that the possessor was actually occupying the land of another.

The possession must be **Exclusive**. The possessor may not share possession with the rightful owner. The possessor must effectively oust the rightful owner from the land. Exclusivity does not mean that the possessor may not share his possession with another person who is not the rightful owner. For example if two brothers who own a farm intentionally fence off their neighbor's property both brothers share in possession against the neighbor.

The possession must be **Continuous** and uninterrupted. The time required for possession varies with the jurisdiction, but it is often between 5 and 20 years. This time is sometimes referred to as the statute of limitations. If the rightful owner is successful in ousting the possessor from the land prior to expiration of the statute of limitations, there is a break in possession and clock is reset to zero. The continuity of possession can be related to the type of use to which the land at issue is commonly put. For example, if the land were only used on a seasonal basis, not occupying the land during the season in which the land would not normally be used may not be an impediment to meeting the continuity requirement.

Many jurisdictions allow **Tacking**. Tacking is being able to use the possession of a predecessor in title in order to extend the possession of the current possessor. A usual requirement of tacking is that the current possessor have **Privity** with the prior possessor. Privity simply means that the present title owner can show an unbroken chain of title to the prior owner. This is usually not a problem when the current possessor has acquired title from the previous possessor. For example, A owns a parcel of land. A erects a fence on his neighbor's property hoping to gain the property through adverse possession. The statute of limitations in the jurisdiction is 20 years. After 15 years A conveys the property to B. If B continues to occupy the property for another 5 years he will have met the 20 year possession requirement. B is in privity with A because B purchased the property from A, and both A and B are in the chain of title.

> Tacking is being able to use the possession of a predecessor in title in order to extend the possession of the current possessor.

Acquiring land through adverse possession seems like stealing the land from your neighbor, doesn't it? Why does the law sanction stealing the land of another? One justification is that owners are required to exercise diligence in keeping their property. The system of law in our country is an adversarial system. If property owners are content to "sit on their rights" for a long period of time and not exercise their legal right to exclude an intruder, they deserve to lose those rights. Another justification is that adverse possession has the effect of settling the location of boundaries, particularly uncertain boundaries. If a property owner is unsure of the location of their boundary they can occupy up to a line of their choice. If it is later determined that the line so established is on the abutting property and the abutter has taken no steps to oust the possessor, the line will have been established through adverse possession. Of course, if the possessor were in a jurisdiction where state of mind was a consideration, for the possessor to be successful he must not have believed that the occupied line was the true boundary.

17.2 Practical Location

In cases where the property owners do not know the location of a boundary, one or both of the abutters may determine a location without resorting to a writing. A number of legal doctrines have developed to support these methods of boundary location. They may be grouped under the general heading of "practical location." The terminology and requirements vary between jurisdictions, but some general categories emerge from the judicial decisions. Because there is no writing to evidence the boundary location or relocation, practical location is an exception to the Statute of Frauds.

17.3 Agreement Followed by Possession

If two abutters are uncertain about the location of their common boundary they may come to an agreement as to its location. If not in writing, the agreement would potentially be in violation of the Statute of Frauds. However, in many jurisdictions a verbal agreement fixing a common boundary followed by possession is considered an exception to the Statute of Frauds.

Depending on the jurisdiction, the parties must either be uncertain about the true location of the boundary, or the boundary must be subject to a dispute. If the parties actually know the location of the common boundary, and agree upon a different location, the agreement is construed as a conveyance. Since a conveyance is governed by the Statute of Frauds, a writing is required. In that case, the verbal agreement would have no effect in changing the common boundary. A similar result occurs if the parties believe they are agreeing to the true boundary, but the boundary agreed upon is later determined not to be the true boundary. Because the parties made a mistake locating the true boundary, they are not prevented from claiming to the true boundary at a later time. The verbal agreement would then have no effect.

The second requirement is possession. The parties must occupy to the agreed line or monument the line in some way. The occupation or monumentation serves as evidence of the location agreed upon. The length of occupation varies with the jurisdiction. In some cases possession must run for the statute of limitations applicable to adverse possession.

17.4 Acquiescence

The doctrine of acquiescence does not involve an actual agreement between the abutters, although an agreement may be implied from the actions of the parties. Acquiescence occurs when one of the abutters erects a fence or establishes other physical evidence of a boundary and occupies up to that boundary. If the other abutter does nothing to dispute the occupation, the line so established may become the boundary through acquiescence. There must be some physical evidence of the claimed line. It could be a fence or a wall or some other clear evidence marking the location.

Acquiescence occurs when one of the abutters erects a fence or establishes other physical evidence of a boundary and occupies up to that boundary.

The length of time required by acquiescence is generally long, often the same period as the statute of limitations for adverse possession. A difference between acquiescence and adverse possession, at least in some jurisdictions, is that in acquiescence an agreement is implied between the abutters. In adverse possession an agreement would destroy hostility and defeat the claim. In acquiescence, although there need be no agreement proved, the fact that one of the abutters did nothing to change the line creates the presumption of an agreement. Silence of the abutter is treated as acceptance of the boundary.

17.5 Estoppel

The term Estoppel means detrimental reliance. One who induces another to act cannot thereafter claim harm from the acts when that party acted in reliance on the first part's representations. For example, suppose an abutter knowing the location of their common boundary, said to a neighbor "go ahead and build your shed right here, it's your property." The neighbor, who doesn't know where the property line is, goes to considerable expense in building the shed. When it is finished, the abutter informs the neighbor that the shed was built on his property and tries to prevent the neighbor from using the shed, preferring to claim it as his own. The abutter will be "estopped" from denying the neighbor the right to use the shed, because the neighbor incurred the expense of building the shed on the abutter's land as a result of the abutter's representations.

The term Estoppel means detrimental reliance. One who induces another to act cannot thereafter claim harm from the acts when that party acted in reliance on the first part's representations.

Here is another example. Suppose a neighbor approaches her abutter telling him she wants to build a wall along their common boundary, and asks if the abutter knows where the line is. The abutter responds, "My boundary line runs between this post and that white oak tree over yonder", pointing out the two objects to the neighbor, "go ahead and build a wall between them." If the neighbor builds the wall just on the neighbor's side of the line, the abutter cannot thereafter claim the wall is on his property. The abutter will be estopped from claiming the line is not where he said it was. The boundary will be established between the post and the white oak tree.

But what if the person making the representations is mistaken about the location of the true boundary, or what if the person who claims to be injured knew where the true line was all along? Many jurisdictions require that in order to be bound by estoppel the person making the representations must have known were the true boundary was located. There is no universal agreement upon this issue, and some jurisdictions do not require knowledge of the true boundary. In that case, the person making the representations would be estopped even without knowledge of the true boundary location. If the person claiming injury knew the location of the true boundary, they could not claim to have relied on the representations of someone else and the person making representations would not be estopped from claiming to the true boundary line.

Chapter 18
Easements

An **Easement** is a right to use the land of another person. Easements are not possessory interests in land. The holder of an easement does not own the land, the holder merely has a right to use the land. An example of an easement is where a land owner grants a power company the right to erect utility poles on their property to provide electrical service to other properties in the area. The land owner still owns the land. However, the ownership is subject to the utility company's rights to use the land. These rights could, for example include the right to drive vehicles over the easement to service the power lines or poles, or to cut brush in order to keep the area accessible. Some easements serve as means of access over land to abutting property. Such an easement is called a right of way.

> The holder of an easement does not own the land, the holder merely has a right to use the land.

In an attempt to clarify easement types, the following discussion groups easements into categories. The reader is cautioned not to place substantial reliance on the names. Different jurisdictions may use different terminology to classify easements. There is often overlap between categories. It is not the name or classification that is important. The legal principles that form the basis of implied easements are often based on equity or fairness, and it is the underlying principles that should be considered.

All easements do not involve unwritten title, but because some of them do, all discussion of easements is included in this chapter for convenience.

P. Gay, *Practical Boundary Surveying*, DOI 10.1007/978-3-319-07158-9_18

18.1 Express Easements

An easement created by a writing is called an **Express Easement**. For example, A grants a parcel of land to B which contains the following language: "said land is conveyed along with a right to drive vehicles over a way 20 feet in width which runs over my remaining land along my westerly boundary from Walden Street to Great Pond…" This language creates an express easement. The easement is called an easement "grant" because the grantor grants the easement to the grantee. A grantor may also convey a portion of her of land and retain a right of passage over the land being conveyed. The language creates a **Reserved Easement**: "…the grantor reserves the right to pass and repass with vehicles or on foot over an old drift way as it is now located on the ground between the highway and the cedar swamp…". It is called a reserved easement, because the grantor reserves the easement to herself out of the parcel being conveyed. By the way, a driftway is a common way or path used for driving cattle.

Easements are interests in land, and are therefore subject to the Statute of Frauds. The easement must be created by virtue of a writing to be effective. As with most legal rules, there are exceptions and sometimes the exceptions swallow the rule. As we shall see, there are a number of ways to create easements without a writing.

18.2 Easement Appurtenant

An easement can be an **Easement Appurtenant** or an **Easement in Gross**. An easement appurtenant, means that the easement serves a parcel of land. An easement appurtenant attaches to the land and passes with title to the land. For example, see Fig. 18.1 "A" owns lot 1 that fronts on Perseverance Way. "B" owns Lot 2 abutting A that has no access to Perseverance Way. A grants B an easement appurtenant over her land so that B can drive her vehicles from Perseverance Way to B's land. B's land is called the **Dominant Estate** or **Dominant Tenement** and A's land is called the **Servient Estate** or **Servient Tenement**.

An easement appurtenant attaches to the land and passes with title to the land.

Easements appurtenant are said to **Run with the Land**. This essentially means that the easement is attached to the land and not to the persons who own the land. Consider the previous example where A granted to B an easement to travel over her land to Perseverance Way. The validity of the easement would not be affected if either A or B conveyed their property. If B sells her land to C, C will have the same rights that B had to access Perseverance Way. However, once B makes the conveyance, B will no longer have any right to use the right of way.

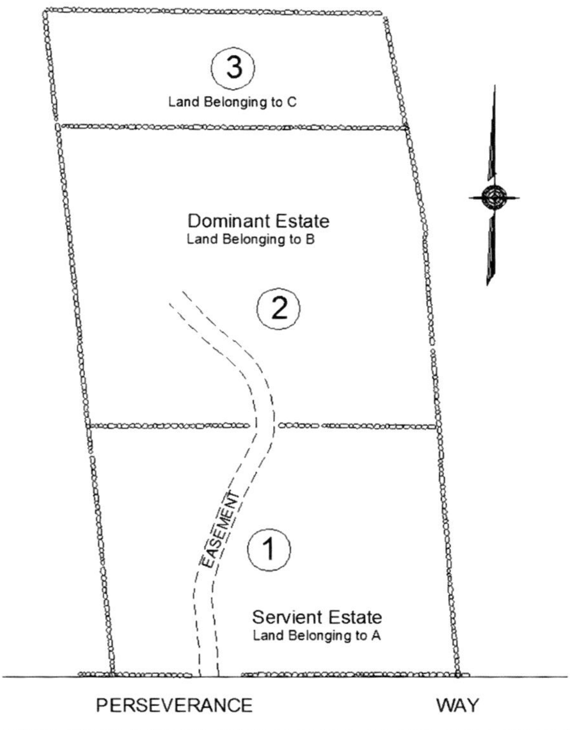

Fig. 18.1 Scope of an easement

Unless there is language in the grant limiting the term of the easement it will run in perpetuity.

18.3 Easement in Gross

An Easement in Gross is an easement that serves a person but does not attach to the land. There is no connection between the easement and the land. In the above example if the easement were in gross, only B would have a right of access to Perseverance Way. If B were to sell her land to C, C would not have a right of travel over A's land. Easements in gross commonly occur where a utility company such as a power company or gas company obtains easements from property owners to run transmission lines or gas pipes over their land.

An Easement in Gross is an easement that serves a person but does not attach to the land.

Where there is no specific language in a grant of an easement as to whether an easement is appurtenant or in gross, the easement is usually construed as being appurtenant, providing that there is land to which the easement may be attached.

18.4 Scope of an Easement

The language which creates a particular easement should state the specific nature of the allowed use of the land. This is often referred to as the **Scope** of the easement. For example, an easement creating a right of way may say that the way is limited to passage by foot only. Or it may say that the way is to be used in the same manner as other ways in town are customarily used. In the first easement, a person can walk over the land but they could not drive a vehicle over the easement. In the second case, if large trucks are normally allowed over the streets in town then they would be allowed to drive over the easement.

An easement is said to be **Overloaded** when used as an easement appurtenant to service land not contemplated in the original easement. Figure 18.1 shows three parcels of land. Lots 1 and 2 were originally in a single common ownership. When lot 2 was sold a right of way was created, by express easement, over lot 1 so that lot 2 would have access to Perseverance Way. The right of way is labeled "Easement" in Fig. 18.1. Sometime later, B, the owner of lot 2 acquires lot 3. Can B use the right of way to service lot 3? The answer is no. To do so would be to extend the use of the easement beyond the original intention of the easement. The easement, as established, is only appurtenant to lot 2. It is not appurtenant to lot 3. Attempting to attach the easement to lot 3 would be "overloading" the easement.

It is also possible to **Overburden** an easement. Overburdening means the easement is being used for uses not intended when the easement was created. Figure 18.2 shows the same two parcels of land and the same easement created in Fig. 18.1. The general rule is that once an easement serves a parcel of land, if the land is later subdivided, the easement serves all of the lots in the new subdivision. In Fig. 18.2, parcel 2 was subdivided into 13 lots. All of the lots will require use of the easement over lot 1 to access Perseverance Way. If there are too many new lots the easement may be overburdened. The issue would be whether or not the additional burden on lot 1 would be considered to be too great and whether the increased use of the easement was beyond that which was originally envisioned when the easement was created.

A common problem facing surveyors in interpreting easements, is where the writing does not specify the location or width of the easement. For example, a right of way running over a 200 acre parcel of land is described in a deed as follows: "including the right to pass and repass from Main Street to the river…" The question facing the surveyor is: where is the right of way and how wide is it? In most jurisdictions, where the location of a right of way is not specified in the grant, the grantor has the right to locate it so long as the location is reasonable. The justification for this approach is that it is the grantor's land that is being burdened by the easement, and it might work a hardship on the grantor if the holder

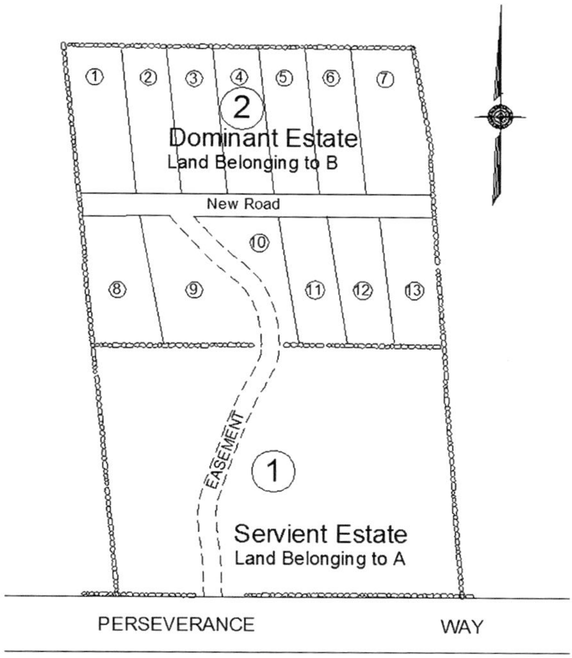

Fig. 18.2 Overburdening an easement

of the dominant tenement were allowed to locate the right of way wherever she desired. As long as the grantor locates the right of way in a reasonably convenient location, the grantee will be required to use that location.

If the grantor fails to locate the right of way, then the grantee may fix the location and width on the ground by continued use. Once the location is fixed by use, neither the grantee nor the grantor can thereafter change the location unless both agree on a new location.

18.5 Prescription

Prescription is the establishment of an easement, such as a right of way, by adverse use. It is an exception to the Statute of Frauds. No writing is required to acquire an easement by prescription. Prescription is much like adverse possession except that in prescription the adverse party acquires no ownership interest in the land. Only a right to use the land is acquired.

Prescription is the establishment of an easement, such as a right of way, by adverse use.

The same elements of adverse possession, actual, open and notorious, hostile and continuous are applicable to establishing prescription. One difference is exclusivity. An easement is by definition the limited use of the land of another person. Unlike adverse possession, the property owners may continue to use their land. So long as the adverse use by the claimant is not interfered with by the property owner, the claim continues to accrue. For example, suppose that A owns land and that A's abutter B, begins walking across A's lawn as a shortcut to a neighbor's house. The weekly practice by A of cutting his lawn over the same location which is used by B as a foot path will not defeat B's claim of prescription. B's use of the land needn't be exclusive. Cutting the lawn would however prevent B from having a valid claim for adverse possession of the footpath. To make that claim, B's possession would have to be exclusive.

In prescription, the location of the use must be in a reasonably fixed area. For example if a person were claiming prescription to establish a right of way over a large field, the user could not move the location of the way to different parts of the field from one year to another.

The scope of an easement by prescription is limited to the type of use which was made during the prescriptive period. For example if a prescriptive easement were created by walking across the abutter's property, the easement would not permit the holder to drive vehicles over the land.

Tacking is also applicable to prescription. In adverse possession, the possessor claims an exclusive occupation of the land. In prescription, the claimant only uses the land. There are judicial decisions which hold that for tacking to be applicable in prescription, the use of the land by the current claimant must be substantially similar to that of the prior user. This requirement would seem to have no applicability in adverse possession because the possession is exclusive so the type of use is usually not a factor.

18.6 Easement Implied from Necessity

The following easements result from **Implication**. Implication means that although the instrument did not specifically call for the easement, it was implied. The language in a conveyance need not make any reference to the easement for it to exist. Circumstances surrounding the conveyance imply the existence of an easement.

An easement Implied from Necessity is created when a common grantor creates a landlocked parcel of land from a larger parcel having access, and the landlocked parcel is conveyed to a person having no other access. For example, suppose the land labeled 1 and 2 in Fig. 18.3 is owned by A as a single parcel of land. The property has frontage on Perseverance Way, a public road. The land is bounded by stone walls and by property belonging to Mr. John A. Butter. There is an unbroken stone wall running east and west across the middle. A conveys the northerly portion of the property to B. The land is now divided into two parcels labeled 1 and 2. The deed makes no mention of a right of way (easement) from parcel 2 over parcel 1 to Perseverance Way. Without an easement over parcel 1, the owner of parcel 2 will have no access to the land.

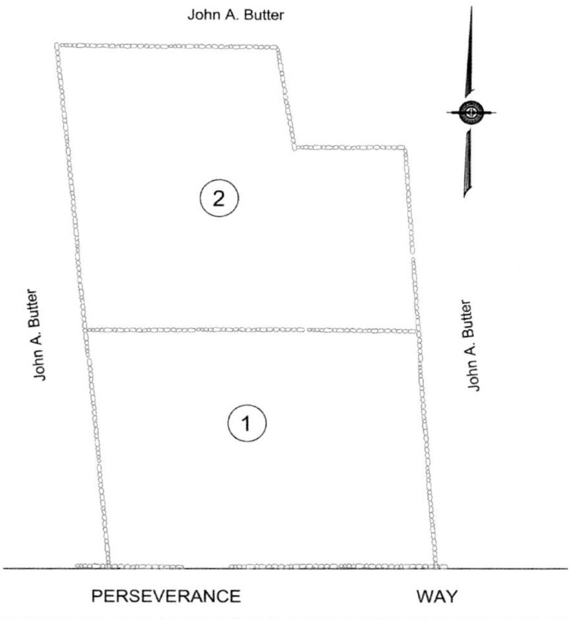

Fig. 18.3 An easement implied from necessity

An easement Implied from Necessity is created when a common grantor creates a landlocked parcel of land from a larger parcel having access, and the landlocked parcel is conveyed to a person having no other access.

Even though there is no express easement for the benefit of B, the law will imply one as soon as the property is conveyed. An easement implied from necessity will then exist over lot 1 for the benefit of lot 2.

No easement would be created if the owner of lot 2 had other access to the property. For example, if parcel 2 were sold to John A. Butter, no easement would be implied because Mr. Butter already has frontage on Perseverance Way and would be able to access his new parcel over his existing land. The lack of access must exist at the time of conveyance of the landlocked parcel otherwise necessity would not exist. Necessity arising after the conveyance would not create an easement by necessity. It should be noted that the necessity usually need only be a "reasonable" necessity. It needn't be an absolute necessity. If Mr. Butter's land did not front on Perseverance Way but only had access to a public road over an old farm road the use of which was limited by wetlands over much of the year the necessity of access over lot 1 might still exist. Although Mr. Butter would technically have access, the access might be so limited as to severely restrict the use of lot 2. In such a case Mr. Butter may have a claim for an easement implied from necessity.

18.7 Easement Implied from Prior Use

An **easement implied from prior** use has some of the attributes of an easement implied from necessity, with the difference that a way exists at the time of the conveyance which apparently provides access to the property being conveyed. Requirements for an easement implied from prior use are:

1. A conveyance of part of the grantor's land.
2. An apparent use of the grantor's land existed at the time of conveyance that provides access to the parcel conveyed.
3. The use is necessary to provide access to the conveyed parcel.

Figure 18.4 illustrates an easement implied from prior use. A owns a single parcel of land completely surrounded by stone walls and bounded on three sides by John A. Butter. A stone wall runs east and west across the land. The wall has a break midway. A gravel driveway runs from Perseverance Way through the break in the wall into the northerly end of the land. A sells the northerly half of the land to B.

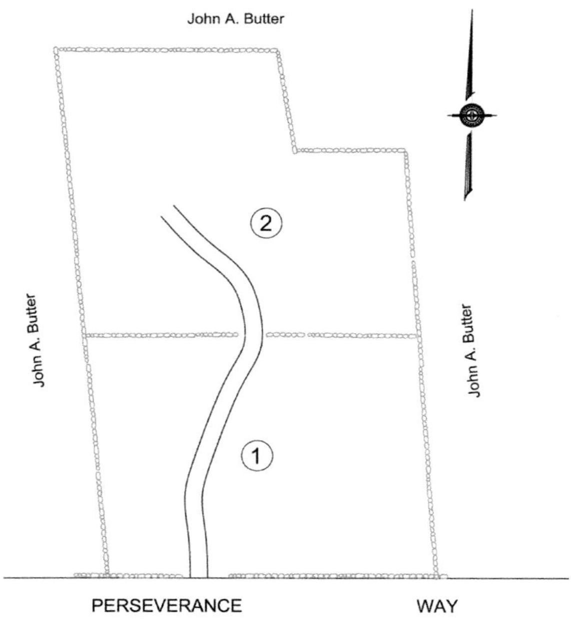

Fig. 18.4 An easement implied from prior use

The deed from A to B describes a parcel of land shown as lot 2 as being bounded on all sides by stone walls. The deed does not mention the gravel driveway and it makes no mention of an easement or right of way to Perseverance Way.

When B viewed the property before purchasing it from A, she drove her vehicle from Perseverance Road along the gravel driveway through the break in the stone wall into the area occupied by lot 2. Although A made no mention of the driveway, it seemed apparent to B that the driveway was the means of accessing lot 2.

In this example, the requirements of an easement implied from prior use have been met. First, the conveyance was a part of the grantor's land. After the conveyance, A retained lot 1. Second, there was a use of the grantor's land that appeared to provide access to the parcel conveyed. A saw and used the gravel driveway to access the parcel from Perseverance Way. Third, the use was necessary to provide access to the conveyed parcel. The parcel conveyed is completely surrounded by stone walls and by the neighbor's land. The gravel driveway provides the only access to parcel 2.

18.8 Easement Implied From a Plat

An **easement implied from a plat** addresses the question of "what rights do owners of lots in a subdivision have to streets in the subdivision, where the streets are not dedicated to the municipality as public ways". The law varies considerably between jurisdictions but three approaches emerge from the judicial decisions.

The first rule, sometimes called the "broad" rule is that all lot owners in a subdivision have a right to use all streets in the subdivision. A second or "intermediate" rule is that a lot owner has a right to use streets in the subdivision that are necessary for the beneficial use of the lot. This is measured by the diminishment of the value of the lot if the use were not allowed. If having the use of a street is necessary to avoid diminishing the lot value, the use is allowed. The third rule is called the "narrow" rule. In this case the only use allowed by the lot owner is that use necessary to provide access to a public street.

The easement implied from a plat only has significance where streets in a subdivision are private, not public. Where streets are public, a lot owner may exercise their right as a member of the public to use any of the streets. Where streets are not public, the theory of easements implied from a plat is applicable.

Many old subdivisions have so-called "paper streets". These are private streets that were shown on the subdivision plan but never built. It often happens that over the years abutters to the streets begin to use the land set aside as a street as their own property. Can the abutters to the streets exclude others in the subdivision from the area platted as a street?

The general rule is that a conveyance of land abutting a street conveys a fee interest to the middle of the street. The fee interest is subject to an easement held by others having rights in the street to be used for purposes for which streets are normally used. It would seem then that in the jurisdictions applying the broad rule, the abutting lot owners would have no right to exclude others in the subdivision. In the intermediate rule jurisdictions an argument of an abutter could be that because the street is not usable as a street—it was never built—others in the subdivision should not be able to claim their lot value would be diminished by being excluded from its use. Others in the subdivision might claim that the street's availability for use adds to the value of their land. The outcome in this case would appear to be far from certain. In the narrow rule states it is unlikely that other lot owners in the subdivision would have a valid claim to the street unless they could show it was their only access to a public street.

Of course, if the abutters to the street were successful in occupying the street area in compliance with adverse possession requirements, they would extinguish the rights of others in the private street.

18.9 Termination of Easements

An easement may be terminated or extinguished. Any easement may be terminated by a written release recorded in the recorder's office. An easement created by necessity will normally be extinguished when the necessity ceases. Let us reconsider the example which applies to Fig. 18.3. In that example A conveyed the northerly portion of the property to B. Under the circumstances an easement implied by necessity was created. If B then conveys the property to Mr. Butter, the easement by necessity will terminate because there is no longer a necessity. Because Mr. Butter's land has access to Perseverance Way and lot 2 is being annexed to the property having access, the necessity ceases.

Easements may be extinguished by adverse acts of the owner of the servient estate. Consider the example shown in Fig. 18.4. Suppose the owner of lot 2 acquires land from Mr. Butter and constructs a driveway over the former Butter land to Perseverance Way. Unlike an easement implied by necessity, the easement by implication would not be extinguished by the alternative access. Suppose further that the owner of lot 1 decides to block off the easement running over lot 1 by building a high concrete wall across the way. If the wall remained in place for the statute of limitations, the easement would be terminated. This theory is really nothing more than the application of adverse possession to extinguish an easement. It should be noted, however, that the blockage of the way must be substantial. Merely erecting a gate or partially blocking the way may not be enough. The blockage must be such that passage is virtually impossible.

It is possible to abandon an easement thereby causing its extinguishment. However, mere nonuse is usually not enough to work an abandonment. There must be some affirmative act on the part of the owner of the dominant estate as evidence of the intent to abandon the easement. For example, in Fig. 18.4, if the owner of lot 2 closed off the wall between lots 1 and 2 so that the way would no longer be passable, the wall would be considered evidence that the owner of lot 2 intended to abandon the easement.

Appendix

A.1 Introduction

The earlier portions of this book intentionally limit the presentation of mathematics in order for the book to be readable by the widest possible audience. Some readers, however, may wish to dig a bit deeper into the underlying theory. If the reader is a student of land surveying, a scientist, engineer, architect or someone with a technical background, then this supplement will provide a more in-depth treatment of the technical principles presented earlier in this book. Because the focus of the book is boundary surveying, this supplement continues in this vein and concentrates on those issues that are relevant to the discipline. Some areas of surveying, such as construction surveying, earthwork, astronomical observations and topographic surveying, for example, are not particularly relevant to boundary surveying so they will not be discussed.

The mathematics presented in this supplement is straightforward and the author has attempted to provide clearly worded explanations of each subject. Although a background in algebra, geometry, trigonometry and statistics would be helpful, none of these is strictly necessary in order for the reader to work through these materials.

Computers and software have made surveying calculations very quick and easy. This does not relieve a land surveyor from understanding the principles underlying a specific task or calculation. On the other hand, most of us lead busy lives so our available time is limited. This supplement therefore takes a pragmatic approach to the subjects. The material presented is what a boundary surveyor should know—what he or she needs to know. We will not spend much time on abstractions and things that are interesting and nice to know, but really not that essential to a working knowledge of the subject. The subjects presented here have been somewhat constrained by limits on the size of the appendix, so, not every subject has been covered. If there are specific areas that, you, the reader, would like to see covered in future editions, the author would love to hear from you.

So, let's get started.

A.2 Angles and Trigonometry

An **angle** is a measurement that describes the geometric relationship between two intersecting lines. Land surveyors in the U.S. use the **sexigesimal system** of angle measurement. Angles are measured in **degrees**, **minutes** and **seconds**. There are 360° (degrees) in a circle, 60′ (minutes) in a degree and 60″ (seconds) in a minute. As can be seen from Fig. A.1, angles are always measured from a reference line and can be measured to the right (clockwise) or to the left (counterclockwise).

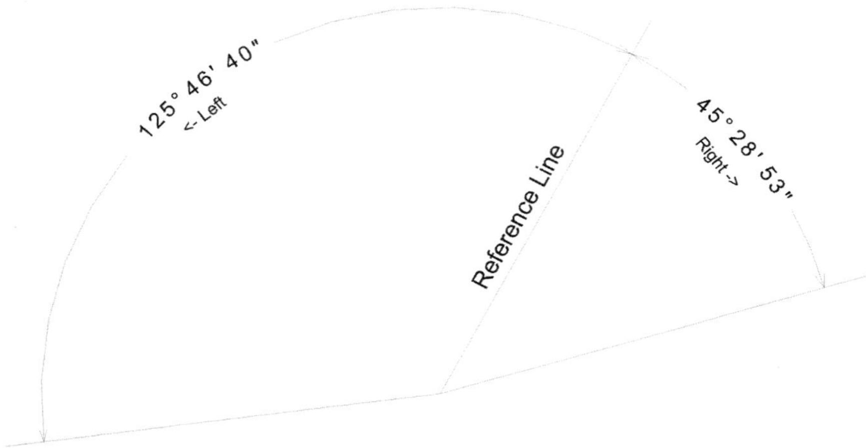

Fig. A.1 Measurement of angles

A.2.1 Adding and Subtracting Angles

When adding and subtracting angles, it is easiest to start with seconds, then minutes, then degrees. In other words, start at the right and move left. For minutes and seconds it is customary to show a maximum value of 59 so if the value is 60 or more, we will subtract 60 and add 1 to the value to the left. For example, let's add the following two angles:

$95° \, 44' \, 32''$ and $10° \, 22' \, 45''$

First we add the seconds: $32'' + 45'' = 77''$. Because this is greater than 59 seconds we subtract 60 seconds: $77'' - 60'' = 1' \, 17''$.

Nest we add the minutes: $44' + 22' + 1' = 67'$. Because this is greater than 59 min we subtract 60 min: $67 - 60 = 1° \, 7'$.

Next we add the degrees: $95° + 10° + 1° = 106°$.

Our answer is: $106° \, 07' \, 17''$.

The above example assumes that the value of seconds and minutes is less than 120. It is entirely possible that the value of minutes or seconds is substantially greater. This often happens when adding many angles together as in the following example:

$$\begin{array}{l} 39° \, 14' \, 45'' \\ 12° \, 02' \, 56'' \\ 19° \, 15' \, 49'' \\ \underline{54° \, 07' \, 39''} \\ 124° \, 38' \, 189'' \end{array}$$

In such cases we just need to recognize that the value is some multiple of 60. In the present example, 189 seconds is greater than 180 seconds ($3 * 60 = 180$), so we would subtract 180 seconds and add 3 min resulting in a sum of $124° \, 41' \, 09''$. We would use the same approach for large values of minutes.

Subtracting angles uses a similar approach. Again, we start from the right. Let's perform the following subtraction:

$102° \, 12' \, 22'' - 11° \, 23' \, 36''$

First we subtract 36″ from 24″. Because we do not want to end up with a negative number we first need to borrow 60 seconds from the minute column. This will decrease the minutes by one.

$22'' + 60'' = 82''$ so, $82'' - 36'' = 46''$

Next we subtract the minutes. Instead of 12′ we now have 11′. Because 23′ is greater than 11′ we need to borrow 60′ from the degrees column and reduce the degrees by one.

$11' + 60' = 71'$ so, $71' - 23' = 48'$
Next we subtract the degrees. $101° - 11° = 90°$
Our answer is: 90° 48′ 46″

A.2.2 *Converting DMS to Decimal Degrees*

Most Scientific calculators require decimal degrees in order to produce trigonometric functions. These calculators are usually able to convert degrees-minutes-seconds (DMS) to decimal degrees and to reverse the procedure. Sometimes the function is listed as converting hours-minutes-seconds to decimal hours. If you need to do it on a calculator without a conversion function, you can easily do so. Consider the angle: 34° 29′ 45″. Start with the seconds. We want to convert the seconds into decimal minutes. There are 60 seconds in a minute so:

$$45'' = \frac{45}{60} = 0.750 \text{ min}$$

Next we add the decimal minutes to the minutes and divide the result by 60 to get decimal degrees:

$$Decimal\ portion\ of\ degrees = \frac{29' + 0.750'}{60} = 0.4958°$$

Finally, we add the decimal portion of the degrees to the degrees to get decimal degrees.

$$Decimal\ degrees = 34.4958°$$

A.2.3 *Converting Decimal Degrees to DMS*

Suppose you have decimal degrees and want to convert it to degrees, minutes and seconds. Using the same angle as above, let's convert 34.4958° into DMS. First, multiply the decimal part by 60 to get minutes:

$$Minutes = 0.4958 * 60 = 29.748'$$

We now have 29 min and need the seconds. So, multiply the decimal part of the minutes by 60 to get seconds.

$$Seconds = 0.748 * 60 = 44.88'' \textit{ rounded off to } 45''$$

Our result is 34° 29′ 45″.

A.2.4 Trigonometric Functions

Trigonometric functions enable us to calculate relationships using angles. The best way to understand these functions is by means of the **unit circle** shown in Fig. A.2. A unit circle has a radius of 1 unit. It doesn't matter what the units are. It could be 1 foot or 1 m. If we draw a line between the center of the circle and the circumference it will have a length of 1. In the figure, we measure the angle of the line from the horizontal axis in a counterclockwise direction. In this case the angle is 60°. If we draw a vertical line from the end of our 1 unit line down to the horizontal X axis, and accurately measured this line it would have a length of 0.87. We call this value a **sine**. If we measure from the center of the circle to this line along the horizontal axis it will measure 0.50. We call this value a **cosine**. Sine is abbreviated sin and cosine is abbreviated cos.

As we rotate our unit line, the sines and cosines will obviously change. Figure A.3 shows how rotating our line to an angle of 35° changes the sine to 0.57 and the cosine to 0.82.

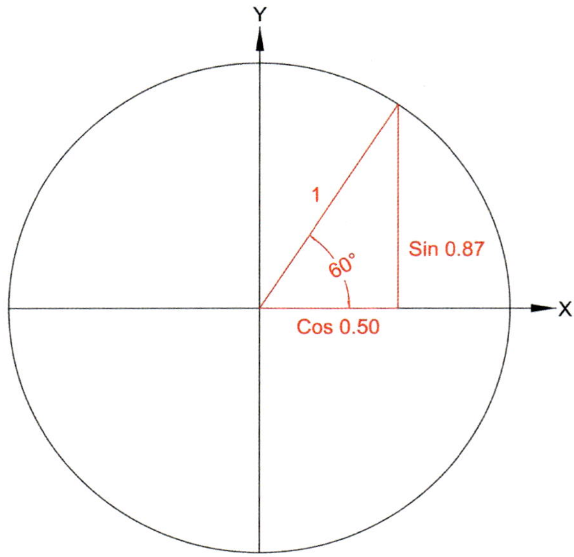

Fig. A.2 Sin and cosine

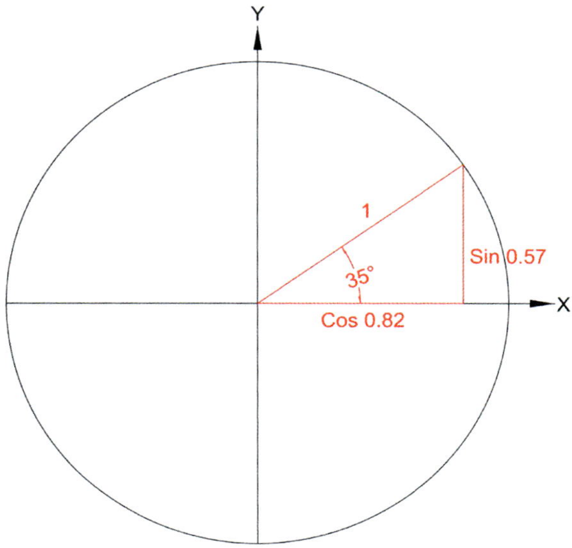

Fig. A.3 Sin and cosine

If we were to rotate our unit line to the left so that the angle became 90° and the line was vertical; the sine would be 1 and the cosine 0. If we were to rotate it to the right until the angle became 0 and the line was horizontal, the sine would be 0 and the cosine 1. We can therefore see from our unit circle that both the sine and cosine will depend on the angle and will vary between 0 and 1. We can also see from this example that if the angle is 45° the sine and cosine will be the same number (0.7071). An additional function to know is the tangent (tan) which is simply the sin divided by the cosine.

So, of what value is this? If we were to use a circle with a radius of any real number, we could use the sins and cosines derived above as factors to calculate the X and Y values. Using the example in Fig. A.3, if our unit line were actually 325 ft. long, the X value would be:

$$325 * \sin 35^\circ = 186.41\text{ ft.}$$

And our Y value would be:

$$325 * \cos 35^\circ = 266.22\text{ ft.}$$

We have seen that every angle has a unique sine and cosine. This means that every sine and cosine must have a unique angle. If the sine of 35° is 0.57, then the angle associated with the sine value of 0.57 must be 35°. In order to differentiate the two cases we say that the arcsine of 0.57 is 35° and the arccosine of 0.82 is 35°. These are abbreviated asin and acos. The exponential notation is a more common way to express these values. So, arcsine is written $\sin^{-1}$ and arcos is written $\cos^{-1}$.

A.2.5 Right Triangles

A triangle is a geometric figure having three sides and three interior angles. A right triangle means that one of the interior angles is 90°. In any triangle, the sum of the three interior angles must equal 180°. So, in the case of a right triangle where one of the angles is 90°, the sum of the other two must also equal 90°.

The most basic, and probably the most commonly used trigonometric solution, in surveying, is a right triangle. Right triangles form the basis of coordinate geometry, which is commonly used in boundary surveying calculations. Right triangles are also used to calculate horizontal distances from slope distances. The reader is encouraged to memorize the three solutions and become adept at solving right triangles. We used right triangles in our unit circles above to demonstrate how sins and cosines are derived.

A right triangle is shown in Fig. A.4. The convention is that angles are noted with upper case letters and the length of the sides with lower case letters. Angle C is 90°. The sides are sometimes given names. The **hypotenuse** is opposite the 90° angle. It is always the longest side. The side next to angle θ is called **adjacent**. The side opposite angle θ is called, believe it or not, **opposite**.

The following three equations will allow you to solve any right triangle:

$$\sin A = \frac{a}{c} \tag{A.1}$$

$$\cos A = \frac{b}{c} \tag{A.2}$$

$$\tan A = \frac{a}{b} \tag{A.3}$$

The above equations can be rearranged as necessary in order to provide the needed solution. If any two of the variables are given, a right triangle can be solved.

Let us look at a few examples of solutions. One common solution illustrated in Fig. A.5, is when a slope distance is measured with a total station and we need the

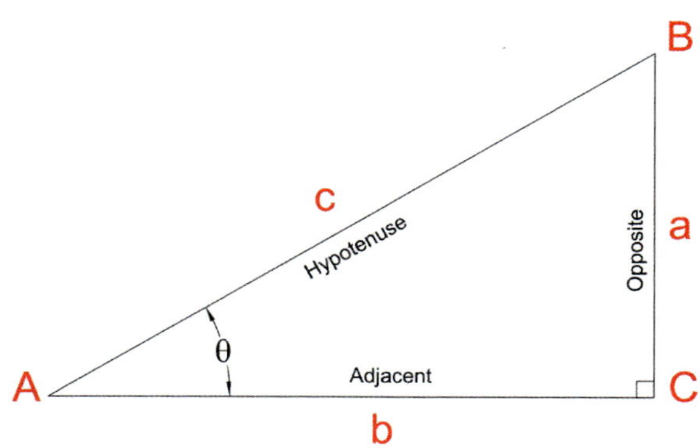

Fig. A.4 Right triangle

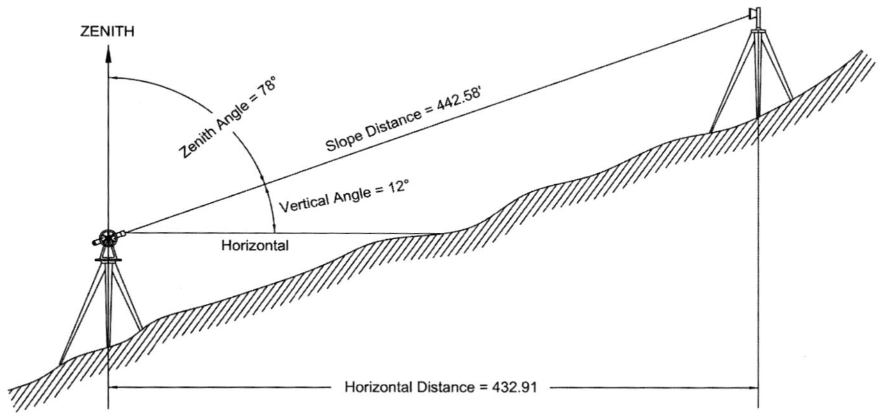

Fig. A.5 Slope distance measured with an EDM

horizontal distance. In addition to the slope distance, the total station provides us with a vertical angle. Normally vertical angles are measured from the zenith (zero is directly overhead), so we will need to subtract the vertical angle from 90° to get the angle from the horizontal plane.

Given: Slope Distance = 442.58 and Zenith Angle = 78°

First, we calculate the vertical angle from horizontal: 90° − 78° = 12°.

Looking at Figs. A.5 and A.6, the slope distance would be side c and the vertical angle, as measured from the horizontal, would be angle A. Imagine the total station at point A measuring to a prism at point B. Line A to C represents the horizontal plane. So, we have c and angle A, and we need distance b, the horizontal distance.

From the equations above we select the one containing all of the three variables: A, b, c:

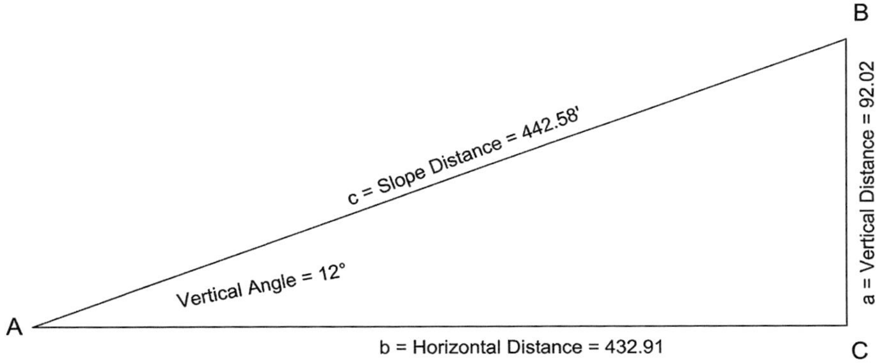

Fig. A.6 Calculate horizontal distance from slope distance

$$\cos A = \frac{b}{c} \tag{A.4}$$

Because we need to solve for b we rearrange the equation as follows:

$$b = c * \cos A$$

Substituting our values in the variables:

$$b = 442.58 * \cos 12^\circ (0.9781) = 432.909\,\text{ft.}$$

Recall that the horizontal distance is always smaller than the slope distance. We could also solve for the vertical distance, which would be side a. Because we now have sides b and c, we could use either Eq. A.1 or Eq. A.3.

Here is a second example. You need to measure the distance from the instrument to a utility pole but the EDM on the total station stopped working. You measure the angle to the top of the pole (15°) and know the height of the pole (30 ft.).

Looking at Fig. A.4, you know A and a. First select the correct equation.

$$\tan A = \frac{a}{b}$$

Rearrange to solve for b.

$$b = \frac{a}{\tan A}$$

$$b = \frac{30}{\tan 15} = \frac{30}{0.268} = 111.96\,\text{ft.}$$

Another useful equation to know when calculating right triangles is:

$$c^2 = a^2 + b^2 \tag{A.5}$$

Knowing the length of any two sides will allow the third side to be calculated. Using Fig. A.6 as an example, assume that we have sides a and b, and we want to solve for side c. We would rearrange Eq. A.5 as follows:

$$c = \sqrt{a^2 + b^2} \quad \text{so} \quad c = \sqrt{92.02^2 + 432.91^2} = 442.58$$

Let's try another example using Fig. A.6. Assume that we know sides c and b and need to calculate side a. Rearrange Eq. A.5 as follows:

$$a^2 = c^2 - b^2$$

$$a = \sqrt{c^2 - b^2}$$

$$a = \sqrt{442.58^2 - 432.91^2} = 92.02$$

A.2.6 Oblique Triangles

Oblique triangles can be more difficult to solve than right triangles. As with right triangles, it is necessary to have three pieces of information in order to solve an oblique triangle. However, with right triangles we already know that one of the angles is 90° so there are really only two unknowns. One possible solution is to break the triangle into two right triangles. Figure A.7 shows an oblique triangle with a line running from B to the base which divides the oblique triangle into two right triangles. However, we shall see that with oblique triangles, it is not always possible to divide and conquer.

The equations governing the solution of oblique triangles, known as the **law of sines**, are:

$$\frac{a}{\sin A} = \frac{b}{\sin B} = \frac{c}{\sin C} \tag{A.6}$$

The law of sines can be used when two angles and one side are known. Let us consider the example shown in Fig. A.8.

Although this triangle could be solved using two right triangles we will use the law of sines. We know angle A, side a, and angle C, so:

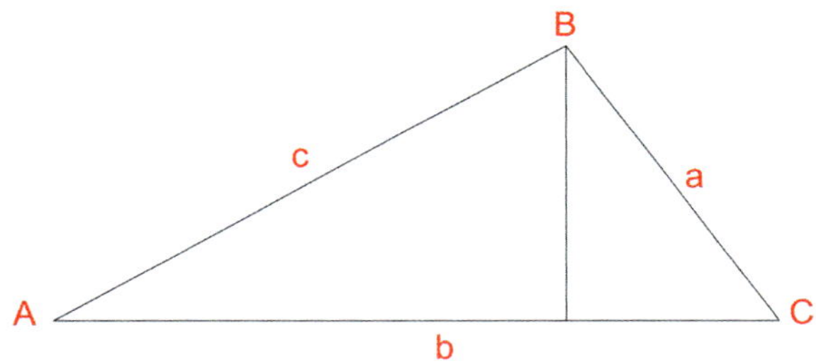

Fig. A.7 Oblique triangle

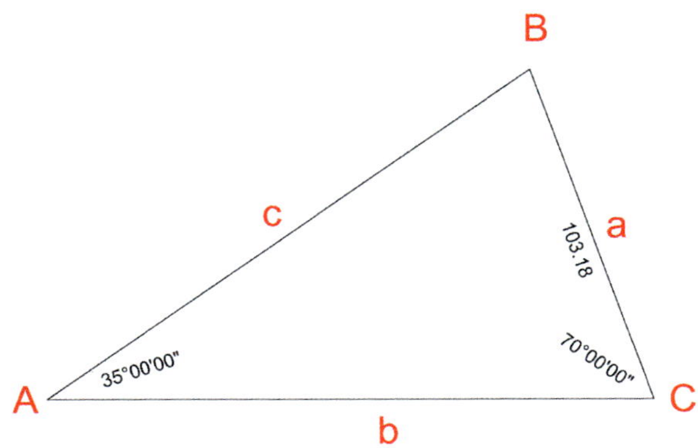

Fig. A.8 Oblique triangle example

$$\frac{a}{\sin A} = \frac{c}{\sin C}$$

Rearranging:

$$c = \frac{a}{\sin A} * \sin C \tag{A.7}$$

$$c = \frac{103.18}{\sin 35^\circ} * \sin 70^\circ = \frac{103.18}{0.574} * 0.940 = 168.97$$

We could then solve for side b by continuing to use the law of sines. Recall that in a triangle, the sum of the interior angles is always 180° so angle B can be easily calculated.

$$B = 180 - (35 + 70) = 75^\circ$$

If we wanted to solve the previous example using two right triangles we could do so by drawing a line from B to a point on side b such that the line would form a 90° angle with side b. We would then have two known variables: Angle C and side a. Using this information we could calculate the length of our newly created line. Once this length was known, we could use angle A and this length to calculate the remaining angles and sides.

Oblique triangles have an ambiguous case where two sides and the angle opposite are known. A number of solutions are possible but they depend on whether the triangle is acute or obtuse. If the angle opposite is acute five possible solutions exist. If the angle opposite is obtuse there are three possible solutions. It is rarely necessary to know how to solve these in boundary surveying, so they are beyond the scope of this chapter. As we will see further on, nearly all of the calculations which need to be performed for the purposes of a boundary survey will be done using coordinate geometry. If the reader wishes to pursue these solutions there are many texts and online resources relating to trigonometry which deal with the subject at length.

A.3 Bearings

Let us first review bearing quadrants. Referring to Fig. A.9, we see there are four bearing quadrants. Each quadrant is named after the two primary directions such as North and East or South and East. Bearing directions are always measured from either the north direction or the south direction. So, as shown in Fig. A.9, a bearing of N15°E is simply an angle measured from north, 15° to the right. N15°W would be in the NW quadrant measured to the left from north. A bearing of S68°E is an angle measured to the left from south.

Bearings are generally expressed as Degrees, Minutes and Seconds as in N34° 22′ 12″E. In order to reverse the direction of a bearing we need only specify the opposite quadrant. For example a bearing having a direction of N22°E can be reversed by relabeling it S22°W.

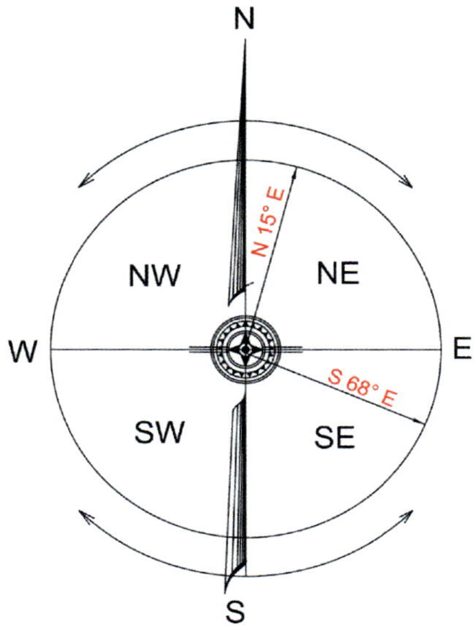

Fig. A.9 Bearing quadrants

The basis of the bearing system in use for any particular project was discussed at length in the main portion of this book. However, as a reminder, bearings can be based on true north, grid north, magnetic north or an assumed north.

A.3.1 Calculating a New Bearing from an Existing Bearing vand an Angle

Now, let us take a look at some calculations using bearings. When calculating a new bearing from a bearing and an angle, we obviously always need a bearing to start with. When working with bearings and angles, there are a number of possible solutions, depending on the magnitudes and of the bearings and angles. We will start with the easiest example first.

In Fig. A.10 we have a starting bearing of N09° 26′ 32″E. We want to add the angle 65° 41′ 58″. The angle is to the right of the bearing. The calculation is straightforward.

$$N09^\circ\ 26'\ 32\mathrm{E} + 65^\circ\ 41'\ 58 = N75^\circ\ 08'\ 30''E$$

Notice that the resulting bearing is less than 90°. If it were greater than 90° it would place the bearing in the SE quadrant.

In the next example shown in Fig. A.11, we start with a bearing of N26° 22′ 34″E and add an angle to the left of 65° 41′ 58″. We notice that the angle is greater than

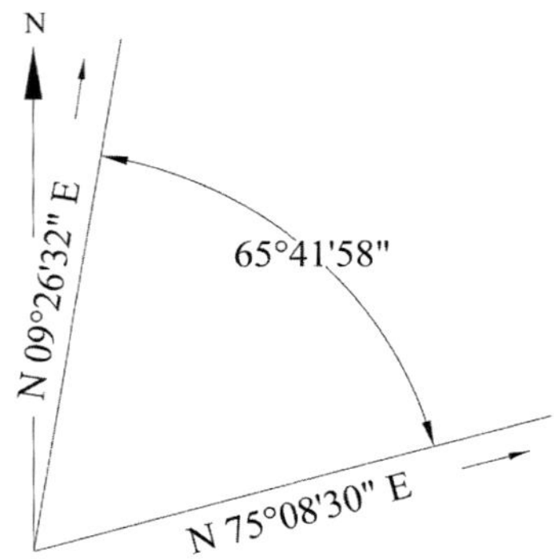

Fig. A.10 Calculate a new bearing from a bearing and angle

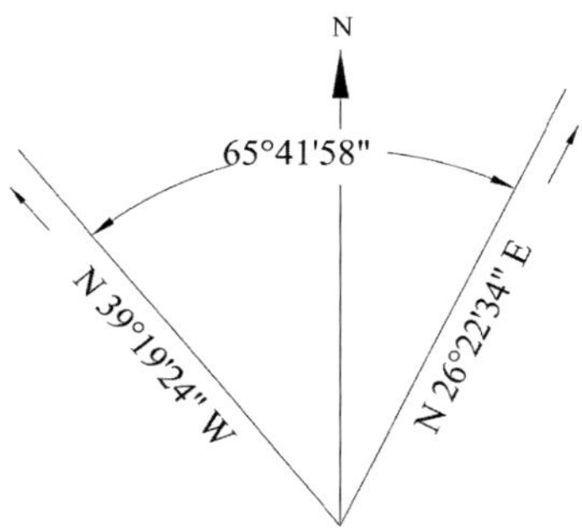

Fig. A.11 Calculate a new bearing from a bearing and angle

the bearing so the new bearing has to be in the NW quadrant. In this case we will want to subtract the 26° 22′ 34″ from the angle in order to get the new bearing.

$$65^\circ\ 41'\ 58 - 26^\circ\ 22'\ 34 = N39^\circ\ 19'\ 24''W$$

The third possible solution is illustrated in Fig. A.12. In this case we start with a bearing of N75° 11′ 39″E and add an angle to the right of 65° 41′ 58″. Adding these together we have:

$$75^\circ\ 11'\ 39'' + 65^\circ\ 41'\ 58 = 140^\circ\ 53'\ 37''$$

We notice that starting from north we have a total angle of 140° 53′ 37″. Because the resulting angle is greater than 90° but less than 180°, we know that the new bearing must be in the SE quadrant. We therefore calculate our bearing by subtracting the angle from 180°:

$$180^\circ - 140^\circ\ 53'\ 38'' = \mathrm{S}39^\circ\ 06'\ 23''\mathrm{E}$$

The next situation is where the new bearing ends up in an opposite quadrant (Fig. A.13). Here we have an initial bearing of S49° 11′ 56″E and an angle to the right of 163° 28′ 28″. We know that the new bearing is in the NW quadrant

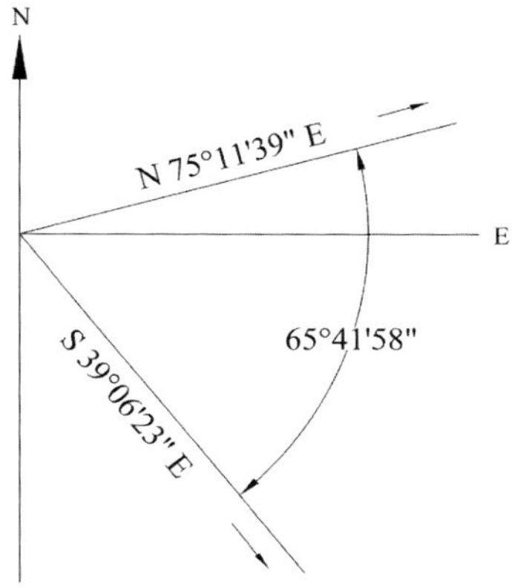

Fig. A.12 Calculate a new bearing from a bearing and angle

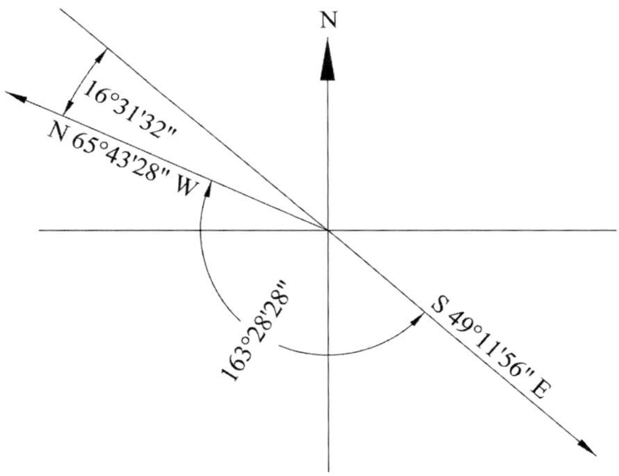

Fig. A.13 Calculate a new bearing from a bearing and angle

because the angle is between 139° and 229°. The 139° is the sum of 90° + 49° 11″ 56″ = 139° (more or less). The number of degrees in the SW quadrant is 90°. The 229° is the sum of 180° + 49° 11′ 56″ = 229° (more or less). The sum of the degrees in the SW and NW quadrants is 180°.

In this example, an easy way to calculate the bearing is to reverse the quadrant of the SE bearing so that it becomes NW. We are simply extending the bearing line into the NW quadrant so that the bearing becomes N49° 11′ 56″W. Because these two bearings are exactly 180° apart, we can subtract the angle and add the difference to the NW bearing.

$$180^\circ - 163^\circ\ 28'\ 28'' = 16^\circ\ 31'\ 32''$$
$$\text{N}49^\circ\ 11'\ 56\text{W} + 16^\circ\ 31'\ 32 = \text{N}65^\circ\ 43'\ 28''\text{W}$$

If you are confused by this procedure, it is always possible to plot the lines on graph paper or draw them in a CAD program until you are comfortable working with the various solutions.

A.3.2 Calculating an Angle from Two Bearings

Here we discuss calculating the angle between two bearings. We can use the same examples we used above.

In Fig. A.14 we are given two bearings in the NE quadrant: N75° 08′ 30″E and N09° 26′ 32″E. The angle is calculated by subtracting the smaller bearing from the larger bearing.

$$75^\circ\ 08'\ 30'' - 09^\circ\ 26'\ 32'' = 65^\circ\ 41'\ 58''$$

Next, we consider two bearings in adjacent N or S quadrants as in Fig. A.15. Here we simply add both bearings together to calculate the angle.

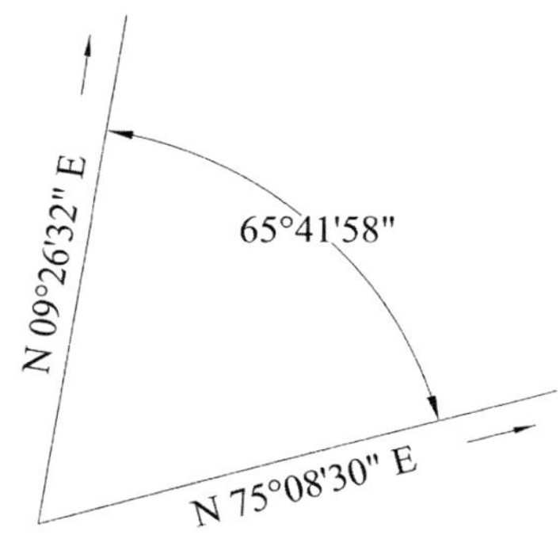

Fig. A.14 Angle to bearing

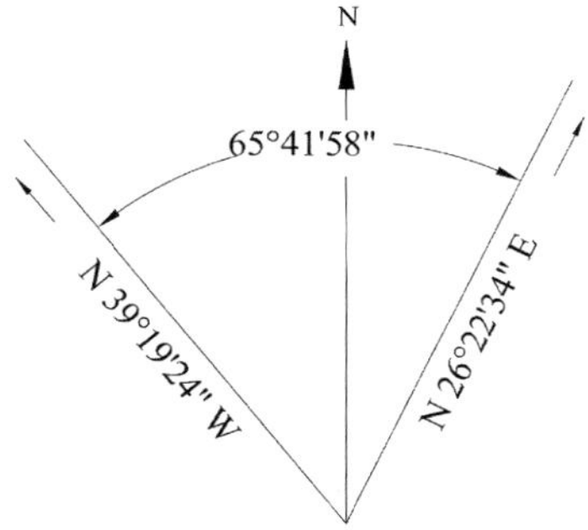

Fig. A.15 Angle to bearing

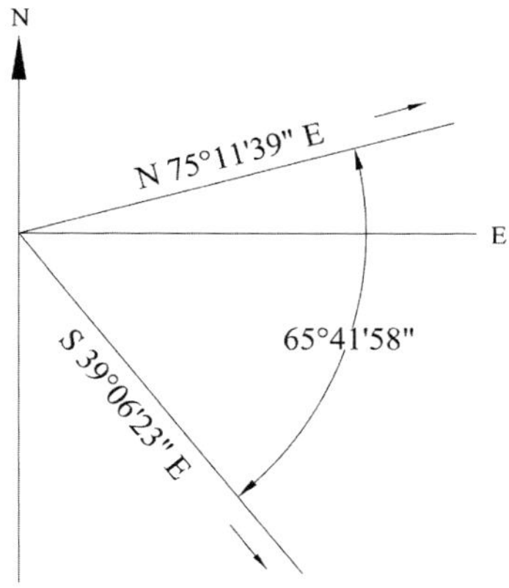

Fig. A.16 Angle to bearing

$$26^\circ\ 22'\ 34 + 39^\circ\ 19'\ 24 = 65^\circ\ 41'\ 58''$$

The third example is in Fig. A.16. Here we need to add both bearings together and subtract the sum from 180°.

$$75^\circ\ 11'\ 39 + 39^\circ\ 06'23 = 114^\circ\ 18'\ 02''$$

$$180^\circ - 114^\circ\ 18'\ 02'' = 65^\circ\ 41'\ 58''$$

Our last example uses two bearings from opposite quadrants in Fig. A.17. We reverse the bearing S49° 11′ 56″E so that it becomes N49° 11′ 56″W. Now we can simply subtract it from the second bearing.

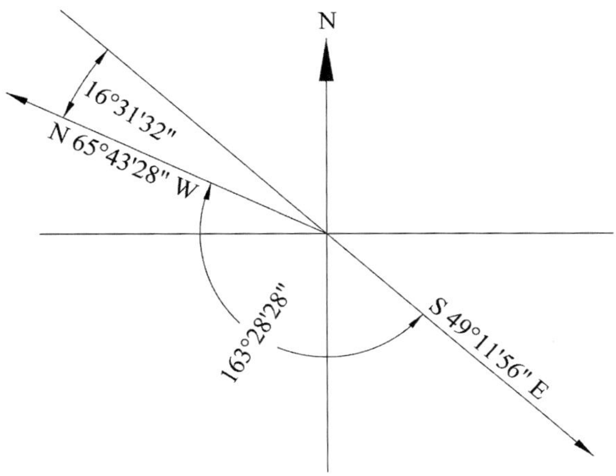

Fig. A.17 Angle to bearing

$$65^\circ\ 43'\ 28'' - 49^\circ\ 11'\ 56'' = 16^\circ\ 31'\ 32''$$

We then subtract the angle from 180° and we have our angle.

$$180^\circ - 16^\circ\ 31'\ 32'' = 163^\circ\ 28'\ 28''$$

A.3.3 Azimuths

Azimuths are angles measured from either north or south. In most cases they are measured clockwise from north, but in some instances they are measured clockwise from south.

As can be seen from Fig. A.18, azimuths are measured from 0° to 360°. There are no quadrants as with bearings. With a bearing, if you need to calculate the reverse direction it is only necessary to change the quadrant designation. The numbers do not change. When using azimuths, it is necessary to add or subtract 180°. The reverse azimuth direction is sometimes called the **back azimuth**. If the azimuth is greater than 180°, subtract 180°. If the azimuth is less than 180°, add 180°. For example, you need to calculate the back azimuth of 89°. It is less than 90°, so you would add 180°, giving 269°.

A.3.4 Deflection Angles

Deflection angles are angles measured from an extension of the previous line.

Deflection angles may be measured to the right or to the left. Figure A.19 shows a deflection angle to the right.

Figure A.20 shows a deflection angle to the left. In both of the preceding images the deflection angles deflect from an extension of line 1–2 and the deflection angle is 15°.

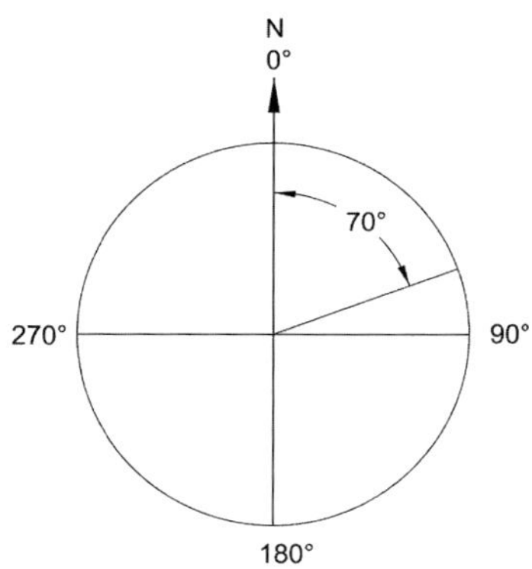

Fig. A.18 Azimuths with 0° at north

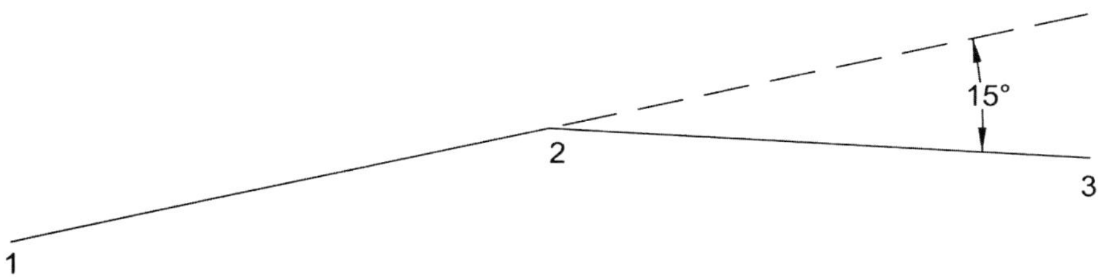

Fig. A.19 Deflection angle to the right

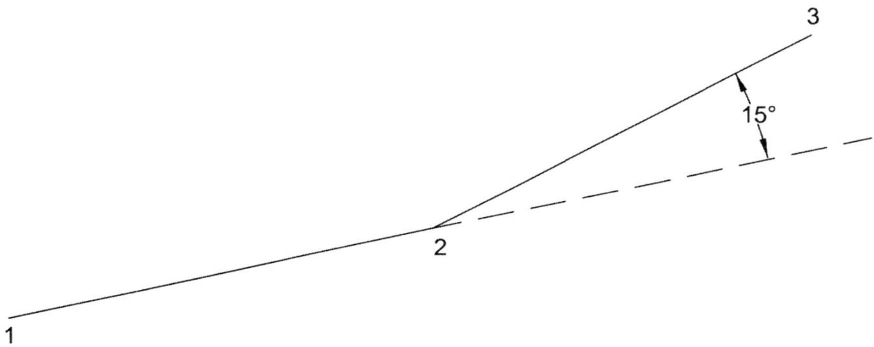

Fig. A.20 Deflection angle to the left

A.3.5 Angular Closure of a Parcel or Traverse

Sometimes we need to check the mathematical correctness of the deed description of a parcel of land described using interior angles. In other cases we need to adjust a closed traverse. In both cases we have a geometric figure having some number of sides and the same number of interior angles. Sometimes, particularly when working with traverses, some of the angles will be exterior and some interior. In this case, we will need to convert all angles to interior angles before proceeding with the angular closure calculation.

In any closed figure, the sum of the interior angles is given by the equation:

$$Sum\ of\ Angles = (n - 2) * 180^\circ \tag{A.8}$$

where n is equal to the number of angles. We have already learned that the sum of the interior angles in a triangle is 180°. We can prove this using the above equation:

$$Sum\ of\ Angles = (3 - 2) * 180^\circ = 180^\circ$$

Let us now check a deed description by summing the angles. As an example, let's say that you plan to perform the boundary survey pictured in Fig. A.21.

The image shows the interior angles that were used to define the boundary lines in a deed description. We can check to see if the angles close using our equation. There are a total of four angles so:

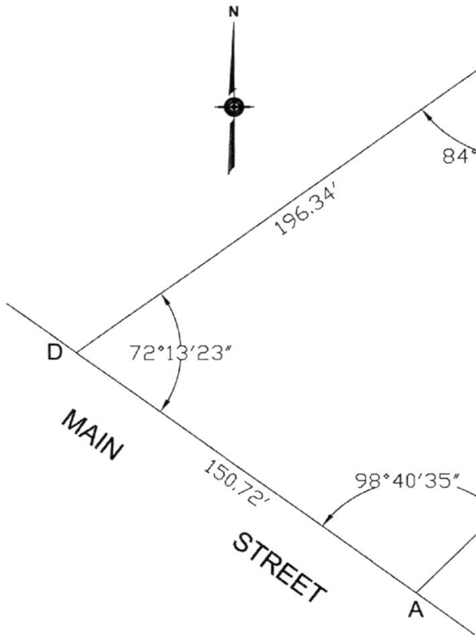

Fig. A.21 Parcel defined by angles

$$\mathit{Sum\ of\ Angles} = (4 - 2) * 180^\circ = 360^\circ$$

The sum of the interior angles in our parcel should total 360°. Starting with angle A and adding the angles we have:

$$\begin{array}{l} 98^\circ\ 40'\ 35'' \\ 104^\circ\ 08'\ 14'' \\ 84^\circ\ 57'\ 48'' \\ \underline{72^\circ\ 13'\ 23''} \\ 360^\circ\ 00'\ 00'' \end{array}$$

So, the sum of the angles is correct and our angular closure is perfect.

When a surveyor runs a traverse, the angles are not likely to close perfectly because, even with the most careful work, there will always be small inconsistencies. Consider the traverse illustrated in Fig. A.22.

There are a total of 7 angles so the sum of the angles should be:

$$\mathit{Sum\ of\ Angles} = (7 - 2) * 180^\circ = 900^\circ$$

We add all the interior angles (Table A.1):

Instead of 900° the sum is 42 seconds off ($60'' - 18'' = 42''$). If the angles closed perfectly, the sum should read 899° 59′ 60″ which is the same as 900°.

Before we can proceed with adjusting our traverse, we should adjust the angles so that they close perfectly. One way to do this would be to distribute the

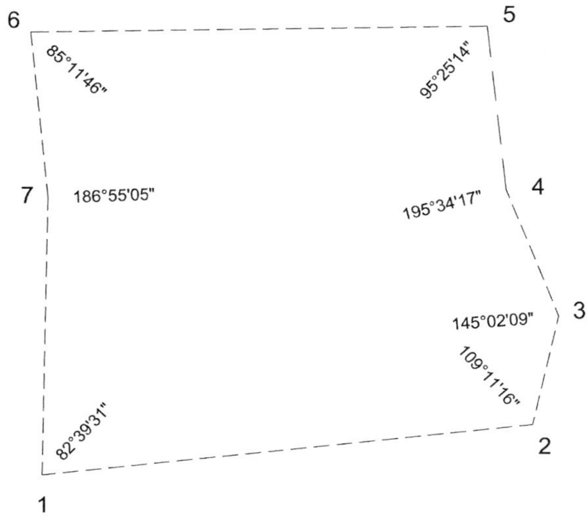

Fig. A.22 Traverse angular closure

Table A.1 Angles as measured in the field

Point	Angle
1	82° 39′ 31″
2	109° 11′ 16″
3	145° 02′ 9″
4	195° 34′ 17″
5	95° 25′ 14″
6	85° 11′ 46″
7	186° 55′ 05″
Sum	899° 59′ 18″

42 seconds error into each angle equally. Another possibility would be to place the error more heavily into one or more of the angles. This is where the experience of the surveyor really comes into play. If experience in the field indicated that one of the setups was on soft ground and there was a high likelihood that the angle was off, the surveyor could go back to the field and measure the angle again. When traversing, surveyors often "double" the angles—turn the angle twice (or even more times) and average the results. This gives greater precision. If the error in a particular set is greater than the other sets, this may be an indication of a lack of precision in the measurement. If so, this angle may be a good candidate for adjustment. Angular errors often occur when the angle is at the intersection of a short line and a long line. For best accuracy and precision, traverses should have the longest lines possible. If the surveyor is relatively certain that the problem is with one particular angle or a couple of angles, the error could be distributed only in these angles, rather than distributing the error equally among all of the angles.

Table A.2 Adjusted angles

Point	Angle	Adjustment	Adjusted angles
1	82° 39′ 31″	06″	82° 39′ 37″
2	109° 11′ 16″	06″	109° 11′ 22″
3	145° 02′ 09″	06″	145° 02′ 15″
4	195° 34′ 17″	06″	195° 34′ 23″
5	95° 25′ 14″	06″	95° 25′ 20″
6	85° 11′ 46″	06″	85° 11′ 52″
7	186° 55′ 05″	06″	186° 55′ 11″
Sum	899° 59′ 18″		900° 00′ 00″

In our example, we will assume that there were no suspect angles and will distribute the error equally into each of the seven angles. So, 42 seconds divided by 7 angles means that the value each angle will need to be increased by 6 seconds. The results are shown in Table A.2.

A.4 Coordinate Geometry

Understanding coordinate geometry is fundamental to performing boundary surveying calculations. During the deed research portion of a survey project, we may want to plot the metes and bounds of deed descriptions. We can use coordinate geometry to do this and to make a determination if the parcel geometry closes mathematically. When we complete our field traverse and have locations of physical evidence, we will use coordinate geometry to plot the points and determine the relationships between record evidence and physical evidence. When we calculate the boundaries of our parcel, we will use coordinate geometry to plot them, perform a lot closure and calculate the area of our parcel.

When using a total station with a data collector, we have the capability of collecting data in three dimensions. To do so however, requires more time during the field portion of the survey because we must constantly keep track of the instrument height and prism height. If using GPS, three dimensional data will automatically be collected. For most boundary surveys, only two dimensional data will be required, because slope distances will be converted to horizontal distances. We will therefore confine our discussion to two dimensional coordinate geometry. Should the need for elevation arise, it can be easily accommodated, but this decision will need to be made prior to starting field work, as instrument and prism heights for every setup will need to be recorded.

A.4.1 The Survey Coordinate System

The first step in beginning our calculations will be to decide on the bearing system to be used. This may have already been decided during the field work phase of our project. For example, if we used GPS to tie into the state plane coordinate system, this will likely be our preferred choice. If we simply used a compass to measure the magnetic direction of one or more of our traverse lines, we will use a magnetic bearing system. If our survey tied into a previously performed survey which made use of a true north bearing system, we may want to adopt that system.

The second decision will be which coordinate system to use. If we are tied into state plane, the decision is easy; we will use the state plane coordinate system. For most small boundary surveys, such as a retracement survey of a house lot it, will usually be more economical just to use an assumed coordinate system.

The coordinate systems we are interested in exploring are based on the Cartesian coordinate system. It is a two dimensional coordinate system having two axes 90° apart. These axes are traditionally labeled as **X** for the horizontal axis and **Y** for the vertical axis. Values increase up and to the right.

In surveying, the Y axis is labeled the **Latitude** (**Northing**) and the X axis the **Departure** (**Easting**). Latitudes increase in a northerly direction. Departures increase in an easterly direction.

Let us consider a simple line as shown in Fig. A.23. The coordinate system is an assumed system beginning with N5,000, E5,000. The actual values of the

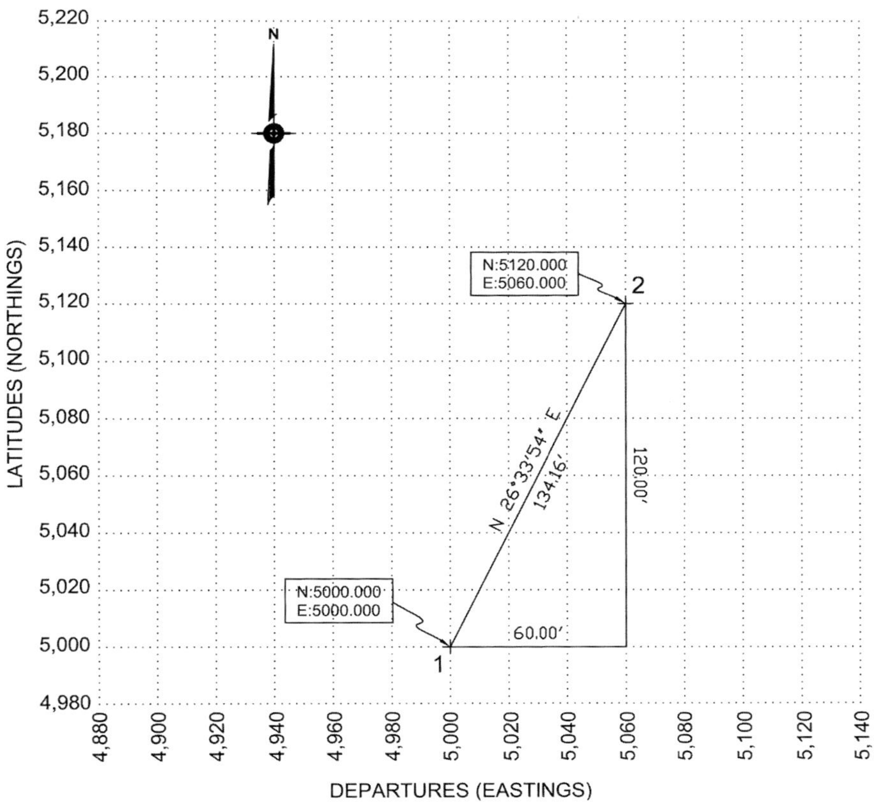

Fig. A.23 Coordinates of a line

starting coordinates do not matter. We could have chosen N100, E100. However, in order to avoid negative coordinate values it is best to pick a large enough starting value, so that all coordinates will remain positive. In a real survey, we may find that, although we thought our field work was nearly complete, we need to extend a traverse down a road 3,000 ft. to the west in order to tie into a monument which has just been discovered. If we had chosen our starting coordinates of N1,000, E1,000 we would end up with negative departure values in the −2,000 range. There is nothing inherently wrong with negative coordinate values, but it is easy to make a mistake by forgetting to include the minus sign. Insuring that all values are positive makes things a bit easier. For small lot surveys, 5,000 is a good choice. For larger tracts you could start with a larger value such as 10,000.

In our example, we have chosen to assign Point 1 the starting value of N5,000, E5,000. Our first (and only) line begins at Point 1 and runs N26° 33′ 54″E, 134.16 ft. to point 2. The purpose of this example is to calculate the coordinates of Point 2.

We have already discussed the importance of right triangles in surveying. Right triangles are an essential part of coordinate geometry. We will use a right triangle to calculate the latitude and departure of our line. This will enable us to calculate coordinates for point 2.

Figure A.23 the latitude and departure of our line. The latitude and departure and the survey line form a right triangle. In Fig. A.24 we have taken the right triangle and transposed it into the familiar orientation for clarity. In this case, we are given side c (134.16 ft.) and angle A (26° 33′ 54″). Notice that angle A was taken from our bearing N26° 33′ 54″E. If we reverse the quadrant of the bearing to the SW quadrant we have S26° 33′ 54″W. Because our 120.00 foot line runs exactly north and south, the bearing is merely the angle between the north–south grid line and our survey line.

We want to solve the triangle for side b and side a. Let's solve for side "a" first. We are given angle A and side c so we will want to use Eq. A.1:

$$\sin A = \frac{a}{c}$$

Rearranging:

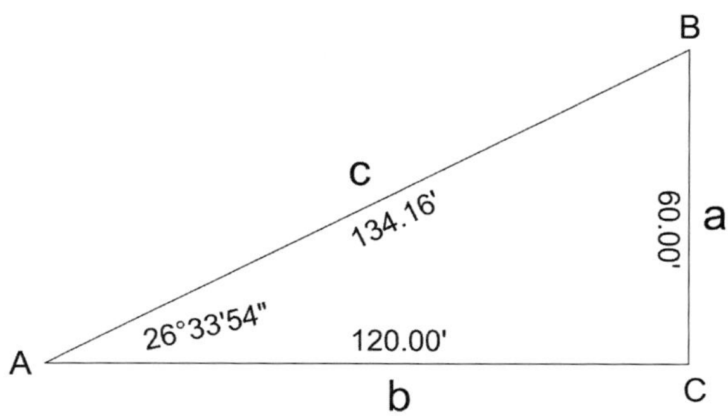

Fig. A.24 Using a right triangle to calculate latitude and departure

$$a = c * \sin A$$

Substituting values:

$$a = 134.16 * \sin(26^\circ\ 33'\ 54'') = 60.00$$

Now, let's solve for side "b". We will use angle "A" and side "c" so the following equation will be used:

$$\cos A = \frac{b}{c}$$

Rearranging:

$$b = c * \cos A$$

Substituting values:

$$b = 134.16 * \cos(26^\circ\ 33'\ 54'') = 120.00$$

If we have a large survey with many courses and distances, using right triangles, as we did above, to calculate latitudes and departures, would be very cumbersome and time consuming. It is much more convenient to perform our coordinate calculations using a tabular format.

Looking at Fig. A.23, we see that side "b" of our right triangle is the latitude and side "a" is the departure. We can use this information to come to the following conclusions:

$$Latitude = Distance * \cos(Bearing)$$
$$Departure = Distance * \sin(Bearing)$$

Using the line shown on Fig. A.23, we can create a table as in Table A.3. In this table, you can see how the start coordinates, the latitude and departure and the end coordinates are calculated and arranged. It should be apparent that using such a table is much simpler and more convenient than figuring out right triangles for each course.

Notice that the latitudes and departures are labeled + and − so that we will know whether to add or subtract them from the coordinate of the previous point. If, as in our example, the line runs North–East, both values will be positive and both the North and East coordinates of the point being calculated will increase. If our line ran North–West, the north coordinate would increase but the East coordinate would decrease.

Table A.3 Latitudes and departures used to calculate coordinates

			Latitude (cos)		Departure (sin)		Coordinates	
Point	Bearing	Distance	North+	South−	East+	West−	N	E
1							5,000.000	5,000.000
	N26° 33′ 54″E	134.160	120.000		60.000			
2							5,120.000	5,060.000

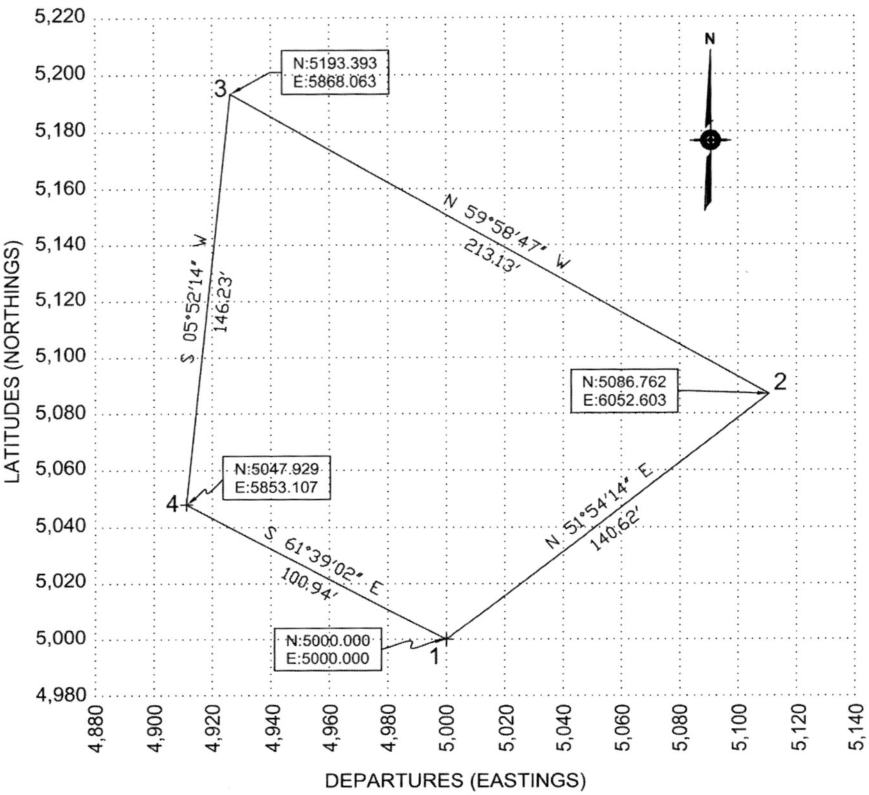

Fig. A.25 A parcel of land with four boundaries

Now that we have an understanding of basic coordinate geometry, let us take a look at a slightly more complex example. The example shown in Fig. A.25 is a parcel of land having 4 boundaries.

The coordinate table is shown in Table A.4. Notice that north latitudes increase the values of the north coordinates and south latitudes decrease their values. Notice that east departures increase the values of east coordinates and west values decrease their values. Also notice that the sums of the north and south latitudes are the same, but with opposite signs to they add up to zero. This is because we started at Point 1 and returned exactly to Point 1. The same is true for the sums of the east and west departures. Also notice that the coordinates for Point 1 are exactly the same at the beginning and end. This tells us that the boundaries of this parcel **close** perfectly. Said in another way, the closure error is zero.

Table A.4 Latitudes and departures

			Latitude (cos)		Departure (sin)		Coordinates	
Point	Bearing	Distance	North+	South−	East+	West−	N	E
1							5,000.000	5,000.000
	N51° 54′ 14″E	140.620	86.762		110.666			
2							5,086.762	5,110.666
	N59° 58′ 47″W	213.130	106.631			−184.540		
3							5,193.393	4,926.126
	S05° 52′ 14″W	146.230		−145.464		−14.956		
4							5,047.929	4,911.170
	S61° 39′ 02E	100.940		−47.929	88.830			
1							5,000.000	5,000.000
		Sums	193.393	−193.393	199.496	−199.496		

A.4.2 Inversing—Calculating the Bearing and Distance Between Two Points

Now that we are comfortable calculating coordinates from bearings and distances, consider the parcel shown in Fig. A.26. Let's assume that we do not know the bearing and distance of the line from 1 to 2, shown as a dashed line. However, we do know the coordinates for points 1 and 2. We need to calculate the bearing and distance between these points. This procedure is called an **inverse**. In our example, we want the direction of the line to start at 2 and run to 1.

We first need to calculate the latitude and departure of the line from the given coordinates. Table A.5 shows a table containing the coordinates of points 1 and 2. Because we want the direction of the line to run from 2 to 1 we need to be aware of whether the coordinates increase or decrease from 2 to 1. The north coordinate for Point 1 is larger than the coordinate for point 2, so the line must run in a northerly direction. The east coordinate for Point 2 is larger than the east coordinate for Point 1 so the line must run in an easterly direction. The plot of the line in Fig. A.26 visually confirms this.

We calculate the values for the latitude and departure by taking the difference between the coordinates. These are recorded in the proper columns in the table, remembering that if the latitude increases it is recorded in the north column. If it decreased, we would record it in the South column. Similarly, if the Departure increases we record it in the East column and if it decreases we put it in the West column. Figure A.27 illustrates the calculation in the form of our familiar right triangle.

Next, we need to calculate the bearing and distance of the unknown line. We know "a" and "b" so we can calculate angle A. Referring to our right triangle equations we choose the following:

$$\tan A = \frac{a}{b}\ or,$$

$$\tan\ Bearing = \frac{Departure}{Latitude} \tag{A.9}$$

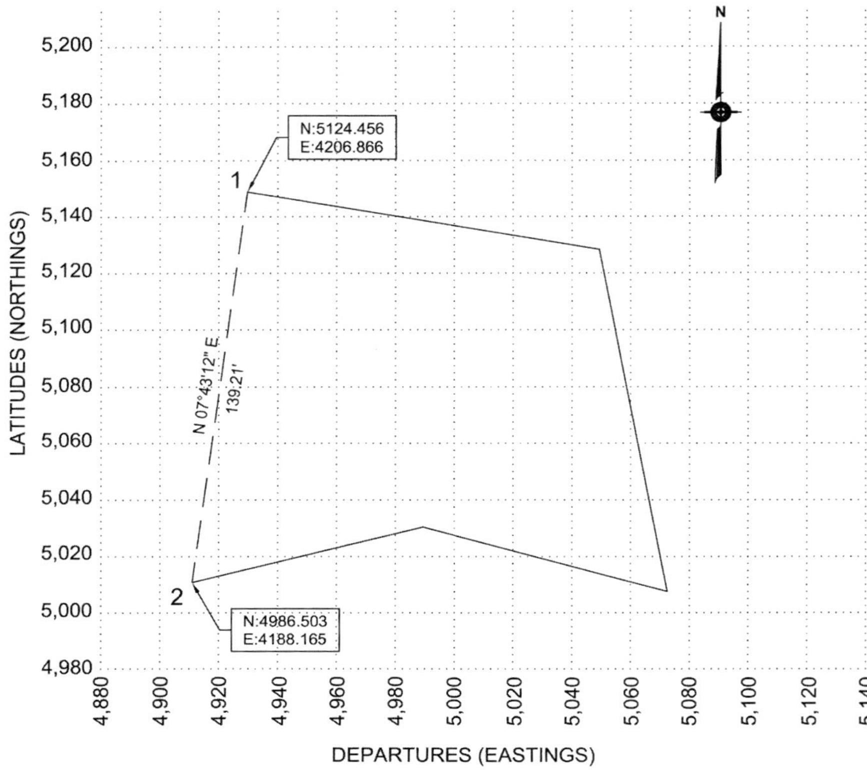

Fig. A.26 An inverse

Table A.5 An inverse

COGO 5 inverse table								
			Latitude (cos)		Departure (sin)		Coordinates	
Point	Bearing	Distance	North+	South−	East+	West−	N	E
2							4,986.503	4,188.165
	N07° 43′ 12″E	139.210	137.953		18.701			
1							5,124.456	4,206.866

Substituting:

$$\tan \textit{Bearing} = \frac{18.701}{137.953} = 0.1356$$

Now that we have the tangent of the bearing, we need the bearing itself, so we use the inverse tangent ($\tan^{-1}$) to find it. Calculators and spreadsheets will return an angle in decimal degrees. In this case the decimal angle is: 7.7200°, which converts to 7° 43′ 12″. We have already determined that the bearing runs northeast from point 2 to point 1 so our bearing is N7° 43′ 12″E.

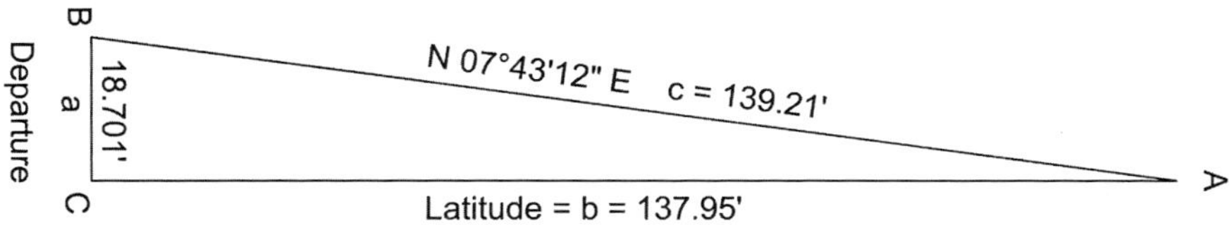

Fig. A.27 Right triangle used to visualize an inverse

Now that we know the angle, it is a simple matter to calculate the distance. From Eq. A.5 recall that:

$$c^2 = a^2 + b^2$$

So:

$$c = \sqrt{a^2 + b^2}$$

Applying the previous equation to latitude and departures:

$$distance = \sqrt{latitude^2 + departure^2} \tag{A.10}$$

Substituting:

$$distance = \sqrt{137.953^2 + 18.701^2} = 139.21$$

Using the methods described above we can calculate coordinates for all of our survey points. We can also calculate the bearing and distance of a line if we know the beginning and end coordinates. If we needed to work with elevation information, this would require adding a third dimension to our coordinate system. The principles used in the calculations would be identical to those already covered. We essentially would take our coordinate grid and rotate it 90° vertically, so that, in addition to an X and Y axis we would have a Z axis.

A.5 Traverse Adjustment

After a traverse has been run in the field it will be necessary to adjust that traverse. Some data collector software will allow this to be done in the field. Many surveyors, however, prefer to perform this phase of the survey in the office where they have access to a desktop computer with a large screen. Offices usually have air conditioning and heat so this is sometimes an added incentive to leave the field in favor of the office.

All traverses will contain some error, no matter how carefully the work was performed. Errors can fall into a number of categories. One category consists of blunders or mistakes. These can be caused by writing wrong numbers into a field book.

Numbers are sometimes transposed or misread. For example 199.98 might mistakenly be written as 198.99. Sometimes measurements are called out over a walkie-talkie and misunderstood. Although relatively uncommon, the memory of a data collector could become corrupted, or a bad cable connection between a data collector and the instrument might cause a corrupted data transfer. Data collectors have substantially reduced blunders compared to the days when every measurement had to be written in a field book. Normal traverse adjustment procedures will not correct blunders.

Other types of errors are systematic errors. This type of error is not random but occurs as a result of some uniform process such as an instrument that is not adjusted properly, a measuring tape that is defective or an instrument person who consistently sights a target incorrectly. Systematic errors are generally not corrected by traverse adjustment.

Random errors can result from a lack of precision in making measurements or in setting up tripods or tribrachs. These errors are usually small and can result in positive or negative errors so that they might cancel each other out. For example, a tripod is set up 0.01′ to the right of point 2. On the next setup the tripod is set up 0.02′ to the left of point 2. Because surveyors work outdoors, sometimes in inclement weather and on difficult terrain, random errors are very common in land surveying. We adjust our traverses and measurements primarily to reduce the effects of random errors.

There are several methods which can be used to adjust a traverse. One is the **Compass Rule**. This method assumes that both the angles and distances were measured with similar precision. The compass rule works well when modern instruments, such as total stations, are used, so the method is widely favored for traverse adjustment.

The **Transit Rule** is another method. This method assumes that angles were measured with greater precision than distances. The **Crandall Method** is another option where the angles are adjusted first and held fixed while the distances are adjusted using a least squares method. It is a relatively time consuming method.

Another method of adjustment is the **Least Squares** method of adjustment. This is a more difficult and time consuming method, and before computers it was used primarily for large and complex control traverses where highly precise results were needed. The advent of computers makes the least squares method more commonly used than it was in years past. Most land surveyors now have software that will perform least squares adjustments, and because it is likely to provide a more refined adjustment it is the preferred method for surveys in which a high order of precision is desired. The least squares adjustment uses the theory of probability to determine the statistically most probable coordinate location for each point in a network. The adjustment provides a statistic best-fit for each point. A least squares adjustment has the additional advantage of providing statistics which tells the surveyor something about the degree of confidence of the calculated position of each point. The surveyor can then accept or reject the adjustment.

A final option is to adjust the traverse based on the surveyor's knowledge of likely sources of error. In this method the error would be placed only in those angles and distances known or believed to have errors. The remaining angles and distances would not be adjusted.

Whatever method is used, the goal is to end up with a traverse that closes perfectly—at least on paper. Because of its long history of use by boundary surveyors and its relative simplicity, our examples in this book will use the compass rule.

A.5.1 Calculating the Error of Closure

Consider the traverse shown in Fig. A.28. This figure contains the raw data which was measured in the field. Because the following is just an example to illustrate the adjustment of latitudes and departures, the angles were not adjusted. However, if this were a real traverse, it would be best to adjust the angles prior to adjusting the traverse. In this exercise all of the angles have already been converted to bearings.

The data has been compiled in Table A.6 to show the latitudes and departures and the coordinates.

Notice that the closing point is numbered Point 8, not point 1. Although the survey began at point 1 and returned to point 1, small errors in the measurements have caused the ending point to be in a slightly different location than the point of

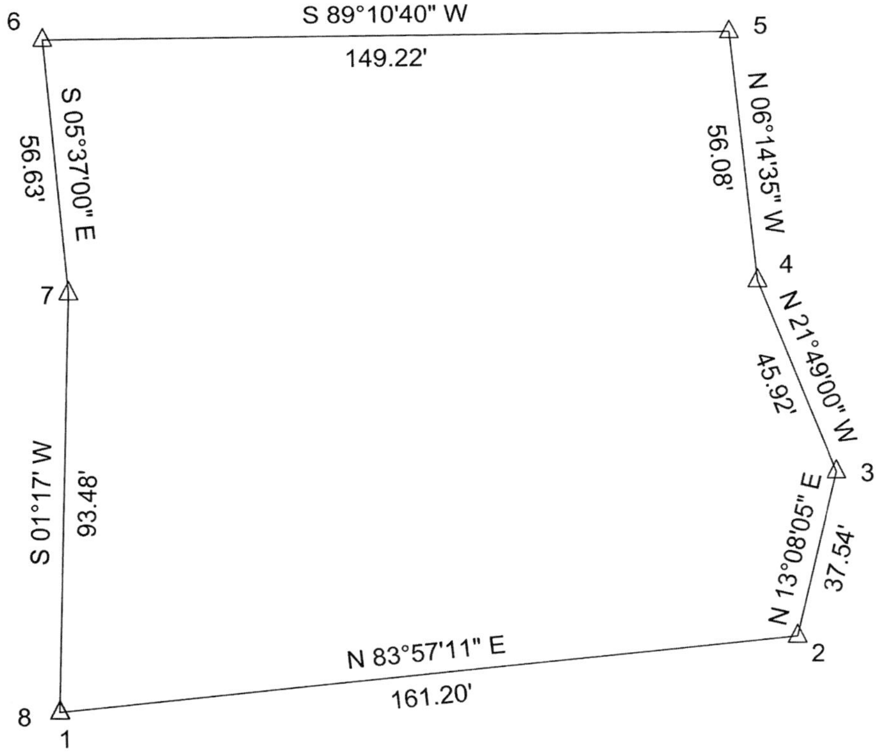

Fig. A.28 Raw data

Table A.6 Traverse raw data table

			Latitude		Departure		Coordinates	
Point	Bearing	Distance	North+	South−	East+	West−	N	E
1							5000.000	5000.000
	N83° 57′ 11″E	161.20	16.981		160.303			
2							5016.981	5160.303
	N13° 08′ 05″E	37.54	36.558		8.531			
3							5053.539	5168.834
	N21° 49′ 00″W	45.92	42.631			−17.066		
4							5096.170	5151.768
	N06° 14′ 35″W	56.08	55.748			−6.098		
5							5151.918	5145.670
	S89 10′ 40″W	149.22		−2.142		−149.205		
6							5149.776	4996.465
	S05 37′ 00″E	56.63		−56.358	5.542			
7							5093.418	5002.007
	S01 17′ 00″W	93.48		−93.456		−2.093		
8							4999.962	4999.914
	Total Dist.	600.07				Error	0.038	0.086
	Sum of Lat and Dep.		151.918	−151.956	174.376	−174.462		
	Difference Lat and Dep.		−0.038		−0.086			

beginning. In order to avoid confusion between the beginning and end points it is usual to assign these points different numbers. Notice that the beginning and ending coordinates are slightly different. This difference represents the error of closure.

The error of closure can be seen by comparing the north and south latitudes (−0.038) and the east and west latitudes (−0.086). This same error is obtained by taking the difference between the starting and ending coordinates. This makes perfect sense because the coordinates are calculated using the latitudes and departures. Comparing the two gives a good check of the math.

We can calculate the bearing and distance of the error by inversing between Point 8 and Point 1. We are already familiar with how to perform an inverse using Eq. A.9:

$$\tan A = \frac{Departure}{Latitude}$$

Substituting:

$$\tan A = \frac{-0.086}{-0.038} = 2.2632 = 66^\circ\ 09'\ 41''$$

As a check, if we look at the coordinates for Point 8 and Point 1 we see that both the north and east coordinates for Point 1 are larger than for Point 8. This means that Point 1 must be northeast of Point 8, so the bearing from Point 8 to Point 1 must be pointing northeast. We also notice that the error of the departure is greater than the error of the latitude so the bearing must be greater than 45° (Fig. A.29).

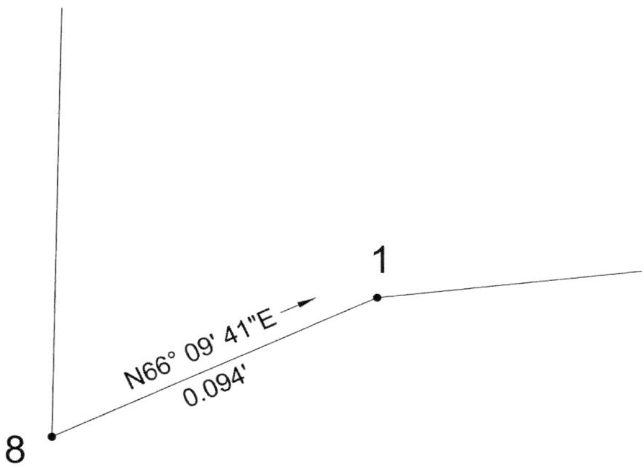

Fig. A.29 Error of closure close-up

Next we calculate the distance between the points using Eq. A.10:

$$distance = \sqrt{latitude^2 + departure^2}$$

Substituting:

$$distance = \sqrt{0.038^2 + 0.086^2} = 0.094'$$

A.5.2 Ratio of the Error of Closure

The error of closure is usually expressed as a ratio of the error to the total distance traversed. The ratio is always expressed with the numerator as 1 so the ratio will be 1 foot in X feet.

$$Ratio \frac{error\ of\ closure}{total\ distance\ traversed}$$

So, in our example the error is 0.094 ft. and the total length of the traverse is 600.07 ft. so, the ratio of Error is:

$$Ratio \frac{0.094}{600.07} = 0.000157$$

Dividing both the numerator and denominator by the error gives the ratio:

$$\frac{1\,foot}{6,383\ \text{ft.}}$$

Hint: after dividing the error by the distance traversed, you can use the 1/x (inverse) key on your calculator to do the division. Sometimes the ratio is expressed to the nearest 100 units so in our case the ratio would be 1 foot in 6,400 ft.

A.5.3 Adjusting the Traverse

The next step is to adjust the traverse. Because the compass rule is one of the most widely used rule for adjusting traverses, we will use it here. We have already calculated the errors in latitude and departure. We will now adjust each latitude and departure so that our traverse closes perfectly.

The compass rule distributes the error to latitudes and departures in proportion to the length of the line. The following equation shows how this is done:

$$Latitude\ Correction\ Line\ X1 - X2 = \frac{Latitude\ Error * Distance}{Total\ Traverse\ Length} \tag{A.11}$$

Using the line from Point 1 to Point 2 as an example:

$$Latitude\ Correction\ Line\ 1 - 2 = \frac{-0.038 * 161.20}{600.07} = -0.010$$

The same procedure is used for departures.

$$Departure\ Correction\ Line\ X1 - X2 = \frac{Departure\ Error * Distance}{Total\ Traverse\ Length} \tag{A.12}$$

The traverse total length is the cumulative length of all traverse lines. In our example the total traverse length is 600.07′.

The adjusted traverse for our example is shown in Table A.7. In order to save space the latitudes and departures have been reduced to a single column each. North and east values are positive and south and west values are negative.

In Table A.7, the Lat. Correction and Dep. Correction columns show the corrections to be made to each latitude and departure in accordance with Eqs. A.11 and A.12. The column of corrections should be added to be sure that the sum is equal to the total correction. For example, the total correction for the departures is −0.038. The sum of the latitude correction column is also −0.038.

After all of the corrections have been calculated, they are applied to the latitudes and departures to create balanced latitudes and departures. Using this procedure, the columns for balanced latitude and balanced departure are completed. The balanced latitudes and balanced departure columns must be added together to ensure that the totals are zero. In other words, for the traverse to close properly the sum of the positive latitudes must equal the sum of the negative latitudes. The same applies to the departures.

After all of the balanced latitudes and departures have been calculated, they are applied to the coordinates. The beginning and end coordinates must be the same. In our example coordinates for Points 8 and 9 are the same. We now have adjusted coordinates for our traverse.

The bearings and distances shown in Table A.7 are no longer correct because the coordinates have changed slightly. We need to calculate new bearings and distances for the traverse lines. This is easily accomplished by inversing between the adjusted coordinates. Our adjusted traverse with the adjusted bearings and distances is shown in Fig. A.30.

Table A.7 Adjusted traverse

Traverse adjusted table

									Corrected coordinates	
Point	Bearing	Distance	Latitude	Departure	Lat. Correction	Dep. Correction	Balanced latitude	Balanced departure	N	E
1									5,000.000	5,000.000
	N83° 57′ 11″E	161.20	16.981	160.303	−0.010	−0.023	16.991	160.326		
2									5,016.991	5,160.326
	N13° 08′ 05″E	37.54	36.558	8.531	−0.002	−0.005	36.560	8.536		
3									5,053.552	5,168.862
	N21° 49′ 00″W	45.92	42.631	−17.066	−0.003	−0.007	42.634	−17.059		
4									5,096.185	5,151.803
	N06° 14′ 35″W	56.08	55.748	−6.098	−0.004	−0.008	55.752	−6.090		
5									5,151.937	5,145.713
	S89 10′ 40″W	149.22	−2.142	−149.205	−0.009	−0.021	−2.133	−149.184		
6									5,149.804	4,996.529
	S05 37′ 00″E	56.63	−56.358	5.542	−0.004	−0.008	−56.354	5.550		
7									5,093.450	5,002.080
	S01 17′ 00″W	93.48	−93.456	−2.093	−0.006	−0.013	−93.450	−2.080		
8									5,000.000	5,000.000
	Sums	600.07	−0.038	−0.086	−0.038	−0.086	0.000	0.000		

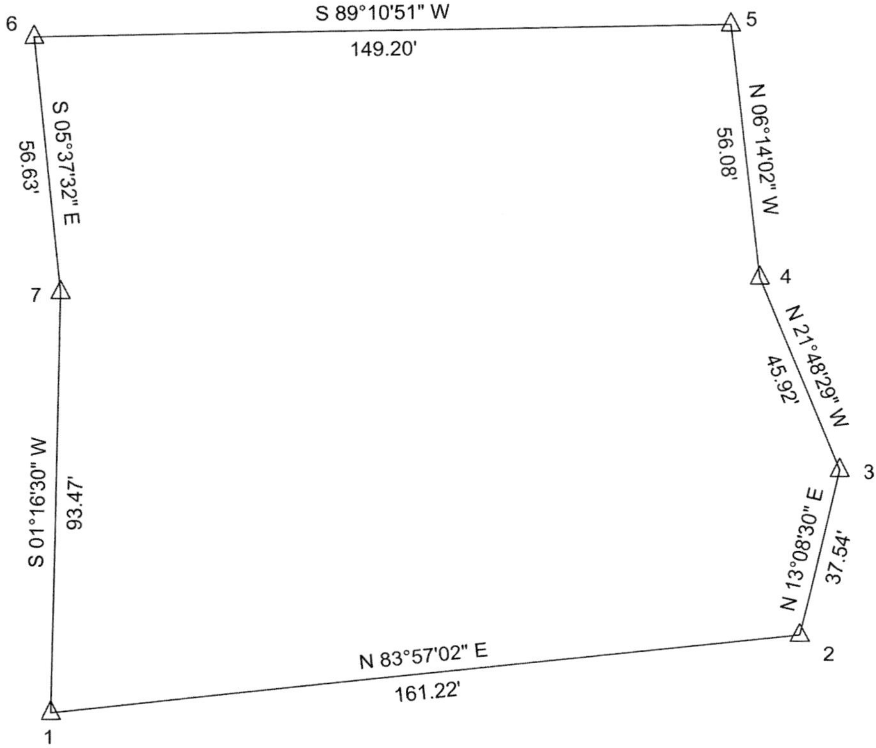

Fig. A.30 Adjusted traverse bearings and distances

A.6 Road Geometry

Most parcels of land have frontage on a public or private road. In this chapter we will review the geometry of roads and road layouts. Calculating road geometry requires a background in plane geometry and trigonometry. We have already covered much of the material that you will need in this section; however some new material on curves will be introduced.

Many roads have **layouts**, sometimes referred to as **alignments**. A layout or alignment describes the curves and tangents of the roadway. Some of these layouts are approved by a government authority such as a local or state government. Some roads have no layouts so determining the location and width of a road or street can sometimes be a challenge for the boundary surveyor.

There may also be applicable standards which govern the design of roads, such as minimum width, minimum curve radii, grade limits, view obstructions and other features. Because our focus is boundary surveying, our primary discussion will concentrate on the two dimensional features of roads. When we use the term **road layout** in this section we will use it in its broadest sense to mean the physical dimensions and geometry of a road. The reader should be aware that the term may also have legal meaning referring to a layout approved by a public agency.

A.6.1 Highway and Road Curves

Most roads are composed of straight sections, called **tangents**, and **curves**. The majority of curves are simple circular arcs which are segments of a circle. Sometimes spiral curves are used, particularly when it is necessary for vehicles to transition from a high speed to a low speed. Spiral curves start from the tangent with gentle curve which gradually increases until a simple arc is reached.

A tangent line to a curve is a straight line that just touches the curve. Figure A.31 shows a circle which has a straight line tangent to it.

If we draw a red dashed line from the center of the circle to a point on the straight line so that the two lines form an angle of 90°, the intersection of the straight lines will be the exact point where the tangent line just touches the circle. The point of tangency of a straight line with a circle or arc will always form a 90° angle between the line and center of the circle or arc.

The nomenclature used to describe various elements of a curve typically used on roads and highways is shown in Fig. A.32. It can be seen from this figure that the road tangents are 90° to the radius lines. Starting on the left side of the figure, the point of tangency is labeled **PC** which is an abbreviation for **point of**

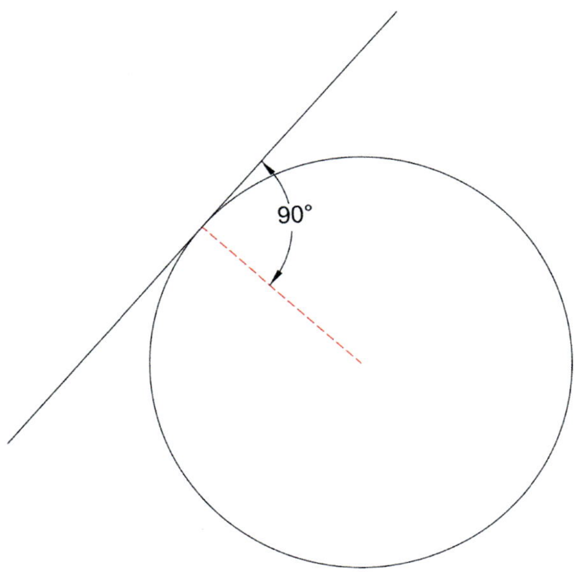

Fig. A.31 Line tangent to curve

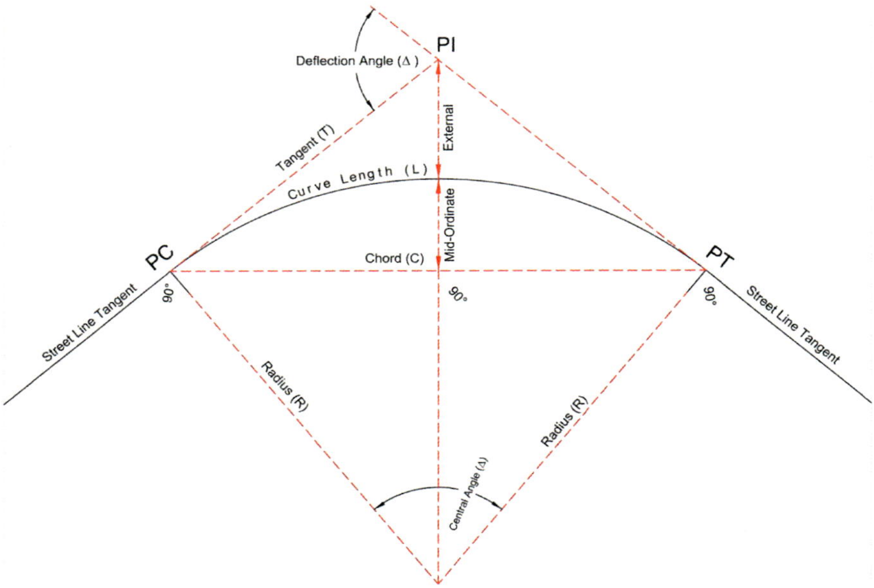

Fig. A.32 Curve nomenclature

curvature. If we follow the curve along to the next point of tangency, we arrive at **PT**, which stands for **point of tangency**. All symmetrical road curves have a PC and a PT. When there is a road layout which begins at one end of the road and traverses to the other end, using bearings for example, the term PC will be used for the beginning of a curve and PT for the end. Think of driving your vehicle down the road in the direction of the layout. You will come to PCs before PTs.

The curve length, abbreviated **L**, is the length of the arc between the PC and PT. The radius (**R**) is the radius of the arc. If a road is designed for slow speed traffic, the radius can be shorter than for high speed traffic. For a given curve angle, as we lengthen the radius, the curve length will increase. Notice the difference in radii and curve lengths between the two examples in Fig. A.33. It would be possible to drive at higher speed along the curve with the larger radius than the one with the smaller radius.

The curve Central Angle (Δ.) is equal to the Deflection Angle (Δ). This can be clearly seen in Fig. A.34.

The Angle between the chord and the Tangent, sometimes called the Tangent Deflection Angle or just Tangent Angle is half the deflection angle. See Fig. A.35.

A.6.2 Calculate Tangent

An example of calculating the length of a Tangent is shown in Fig. A.36. We can see from this figure that the geometry is our familiar right triangle. We have labeled the three angles in the right triangle A, B and C. Recall that in a normal

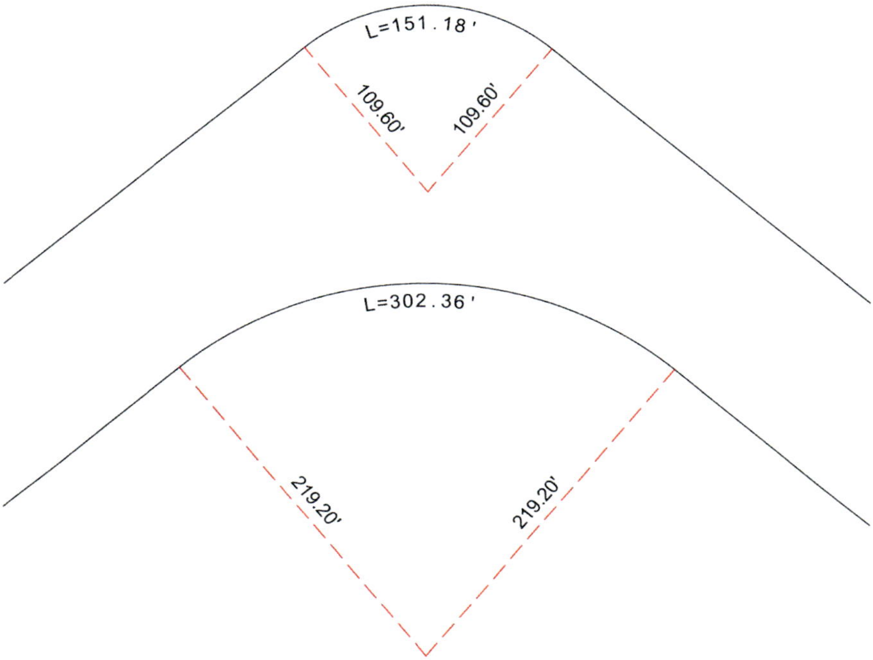

Fig. A.33 Different curve radii

curve the radius line is always tangent to both the street line and the tangent line (the tangent line is simply an extension of the street line).

In the example, we know the curve radius: 275.00′ and the Central Angle: 40° 10′ 00″. We know that the angle between the radius line and the middle ordinate is half the central angle so we calculate angle A of our right triangle to be 20° 05′ 00″. Using Eq. A.3 we rearrange to solve for side a (tangent line).

$$\tan A = \frac{a}{b} \quad \text{so} \quad a = b * \tan A$$
$$a = 275.00' * 0.3656 = 100.545$$

Because angle A is simply $\frac{\Delta}{2}$ we can write an equation for calculating the tangent as follows:

$$T = R * \tan \frac{\Delta}{2} \tag{A.13}$$

A.6.3 Calculate the Length of a Curve

We will next calculate the length of a curve. The curve length is the length of the arc between the two tangent street side lines. We know that a circle contains 360°. Our curve is simply a portion of a circle. The Central Angle (Delta or Δ) tells us

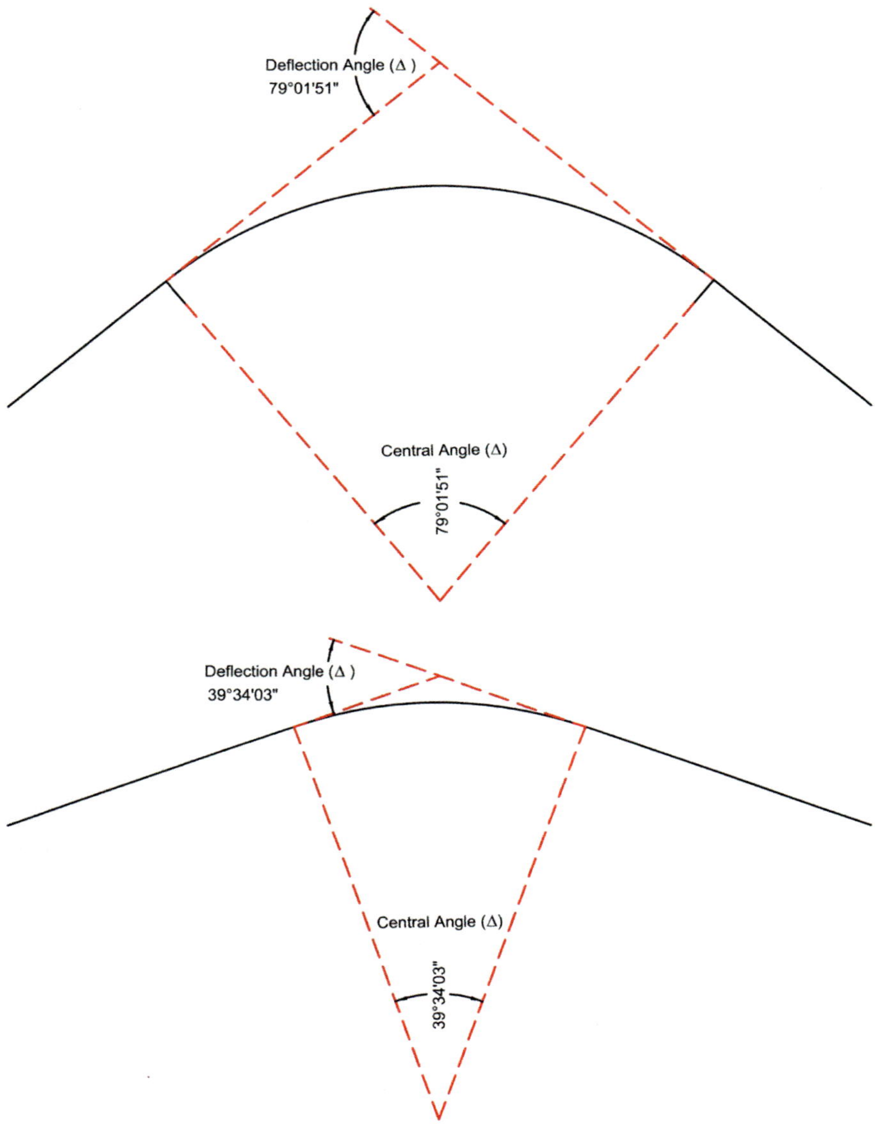

Fig. A.34 Delta angle = central angle

what portion of the circle our curve represents. In the curve shown in Fig. A.37, the central angle is 40° 10′ 00″. We can use a proportion to calculate what fraction of a circle our curve represents. We first need to convert our central angle to decimal degrees then proceed as follows:

$$Fraction = \frac{40.1667^\circ}{360^\circ} = 0.111574$$

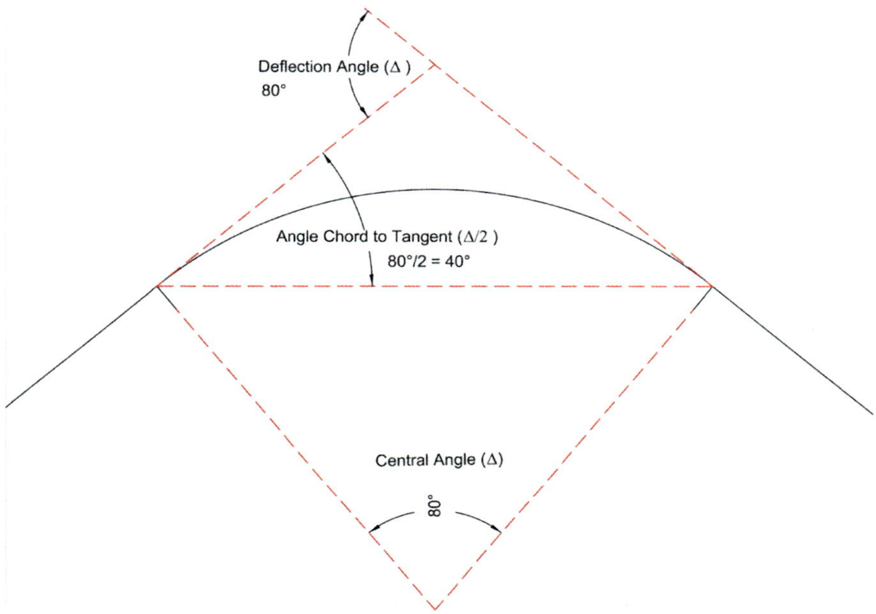

Fig. A.35 Tangent angle

Next we need to know that the circumference of a circle is calculated as:

$$C = \pi * D$$

where C is the circumference of the circle and D is the diameter. In our example the radius is 275.00 ft. so the diameter is 550.00 ft. The circumference of our circle is therefore:

$$C = \pi * 550.00' = 1,727.876'$$

To calculate our curve length we simply multiply the circumference by the fraction of a circle that our curve represents:

$$L = 1,727.876' * 0.111574 = 192.786'$$

We can generalize our equation to calculate the curve length as follows:

$$L = 2 * \pi * R * \frac{\Delta}{360^\circ} \tag{A.14}$$

Using our equation to check the above result:

$$L = 2 * \pi * 275.00' * \frac{40.1667^\circ}{360^\circ} = 192.786'$$

Suppose we only have the radius and the curve length and want know the central angle? We can rearrange Eq. A.14:

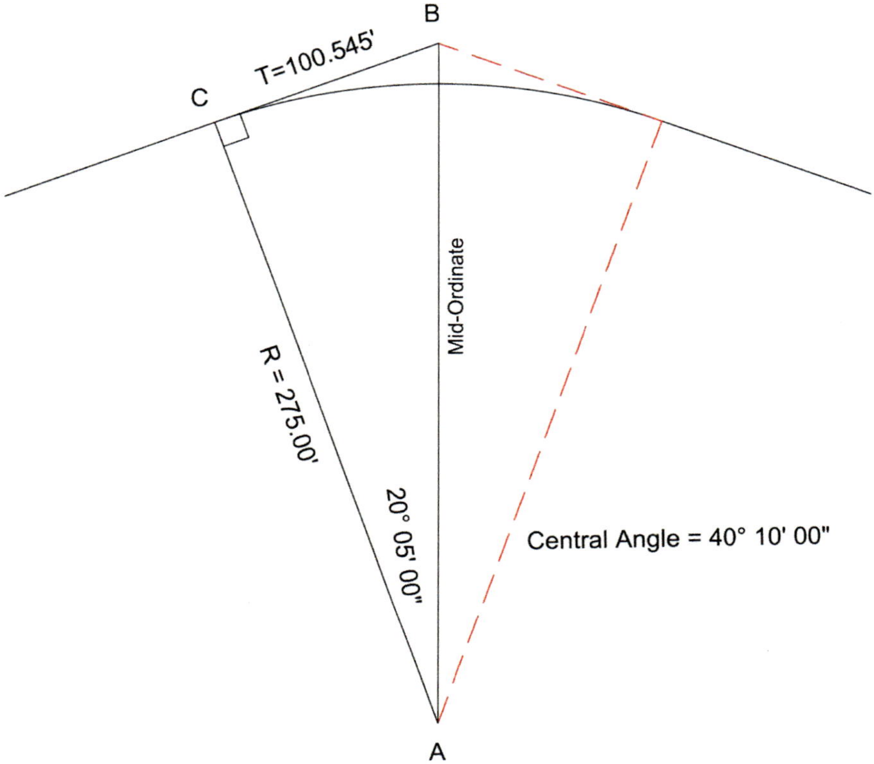

Fig. A.36 Calculate tangent

$$\Delta = \frac{L * 360^\circ}{2 * \pi * R} \tag{A.15}$$

Substituting our values:

$$\Delta = \frac{192.786 * 360}{2 * \pi * 275.00} = 40.166 \textit{ decimal degrees}$$

And converting from decimal degrees to DMS we have: $40^\circ\ 10'\ 00'$.

A.6.4 Curves at Intersections

We will next consider curves typically found at the intersections of streets as shown in Fig. A.38. These curves are really no different than the highway curves which we just examined.

The curve geometry will have a Radius, Delta, Chord and Tangent. In many, if not most, cases it is convenient for the radius to be the same for both sides of the

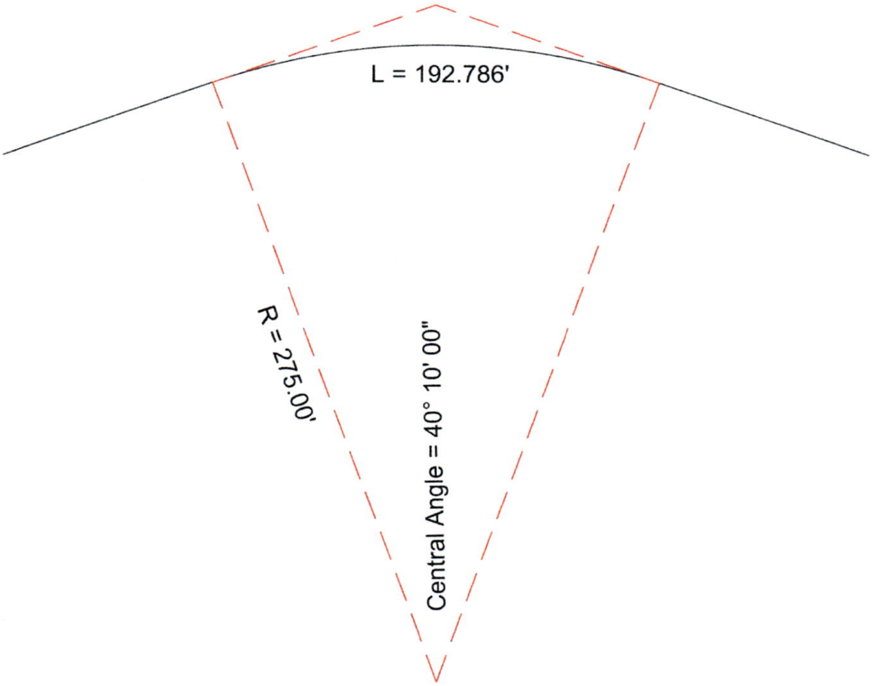

Fig. A.37 Calculate curve length

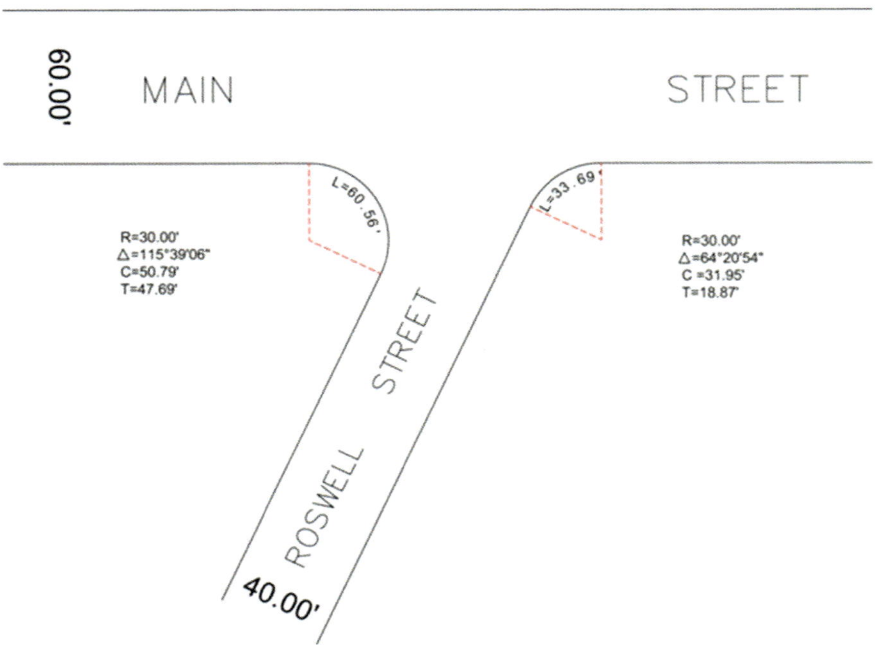

Fig. A.38 Curve at street intersection

intersection, although in some cases it is not. In a typical example, as shown in Fig. A.38, the radius is 30.00 ft. on both sides of Roswell Street. However, it is apparent from the layout that the Delta, Chord and Tangent vary from one side to the other.

In our example, assume that Main Street is straight and that the two sidelines of Roswell St. are parallel. In this geometry, the two Deltas for the curves must add up to 180°. If we do the math and add 115° 39′ 06″ and 64° 20′ 54″ we will get 180°. The remaining values can be calculated using the equations and methods used above.

A.6.5 Compound Curves

A compound curve exists when there are two adjacent curves, each having a different radius. Figure A.39 shows a compound curve.

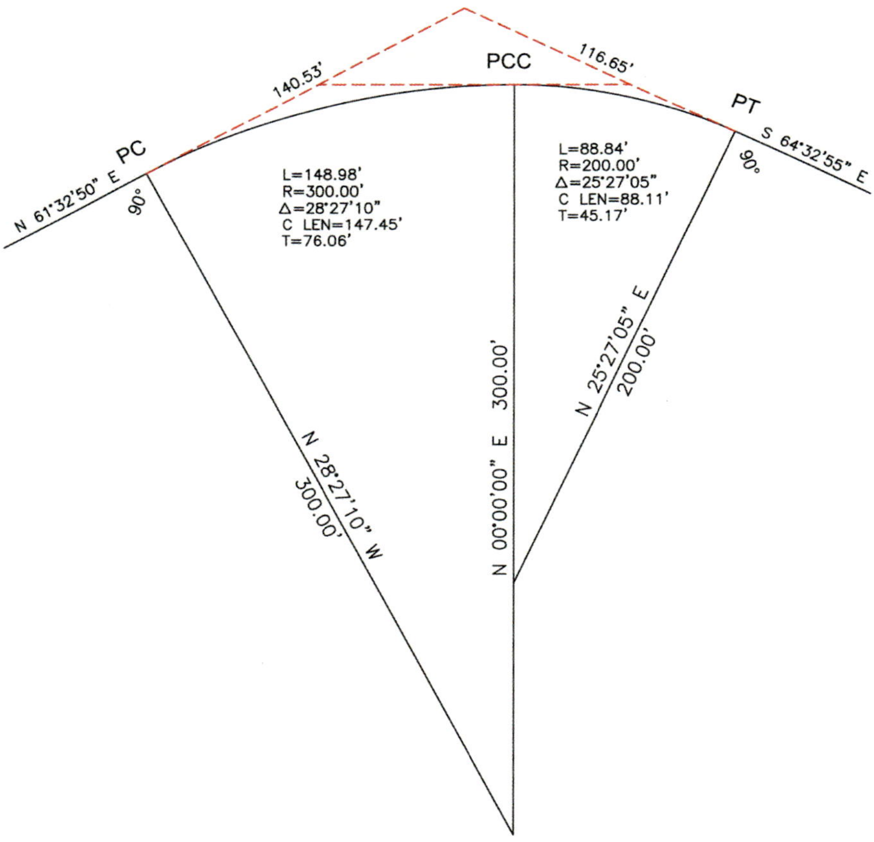

Fig. A.39 Compound curve

Notice that the curve on the left has a radius of 300.00′ and the curve on the right has a radius of 200.00′. Because both curves share a common radius line, the curves are tangent to each other. The methods for calculating the tangents and curve lengths for each of the curves are the same as for the simple curves that have already been discussed.

The intersection of the tangents of the compound curve does not occur opposite the central radius line because of the different radii. Notice that the lengths of the tangents are different. The left side is 140.53′ while the right side is 116.65. One way to calculate the tangent lengths is to use an oblique triangle solution.

In the example shown in Figs. A.39 and A.40, we know the bearings of the tangents. These bearings are simply the bearings of the street lines extended. We also know the bearing of the center radius line and we know that the tangent at PCC (Point of Compound Curve) is at 90° to the radius line bearing. Because we have three bearings we can calculate the interior angles of the oblique triangle. These are shown in Fig. A.40.

We also know that the distance along the base of our triangle is the sum of the lengths of the tangents of each of the curves: 76.06′ + 45.17′ = 121.23′. With this information we can use Eq. A.6 to calculate the unknown side of our triangle. First we combine and rearrange the equations:

$$a = \frac{b}{\sin B} * \sin A$$

Next, we substitute the values and calculate the length of side a.

$$a = \frac{121.23'}{\sin 126^\circ\ 05'\ 45''} * \sin 28^\circ\ 27'\ 10 = 71.48'$$

The total length of the tangent is the sum of the value we just calculated and the value of the tangent of the curve:

$$T = 71.48' + 45.17' = 116.65'$$

We would calculate the remaining tangent using the same method.

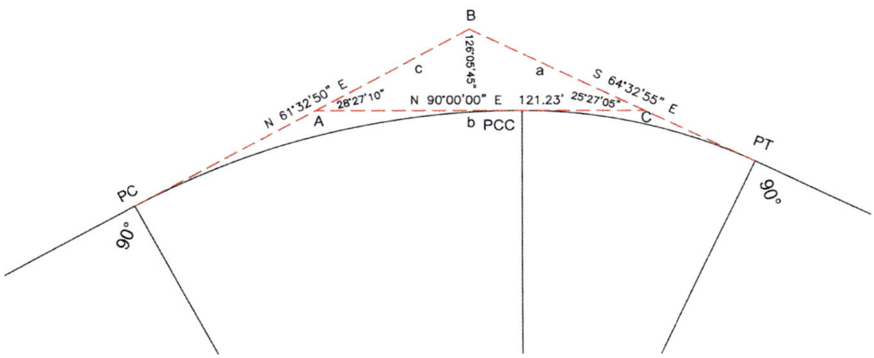

Fig. A.40 Compound curve

A.6.6 Reverse Curves

When a curve immediately changes direction, it is called a reverse curve. Figure A.41 shows a reverse curve. Reverse curves can have identical radii or each curve can have a different radius. The calculations are similar to that of the simple curves that we have already covered, and so do not bear repeating here. In order for both curves to be tangent at the point where the curve changes direction, the radius line of each curve running through Point of Reverse Curve (PRC) must form a straight line. This can be seen in the image where the bearing of each line, N39° 30′ 56″E, is identical.

A.6.7 Road Stations

When designing and laying out roads it is customary to use **Stations.** Stations are commonly numbered using integers which represent 100 ft. So, the distance between Station 5 and Station 9 would be 400.00 ft. When it is necessary to identify a point between stations, it is designated using a + sign. This is illustrated in Fig. A.42. The intersection of the centerlines of Main St. and Roswell Street occurs at Main Street station 10 + 90.91. The intersection at the centerline of Roswell Street occurs at Roswell station 5 + 31.51.

Calculating the distances between stations is straightforward. For example calculate the distance between station 9 and station 10 + 90.91.

$$\begin{aligned} \textit{Station}\,9 &= 9 * 100' = 900' \\ \textit{Station}\,10 + 90.91 &= \left(10 * 100'\right) + 90.91' = 1,090.91' \\ \textit{Distance} &= 1,090.91' - 900' = 190.91' \end{aligned}$$

A.6.8 Road Baselines and Centerlines

Stations can be on a **Baseline** or **Centerline**. Both streets in Fig. A.42 have centerlines. Centerlines are equidistant from the street sidelines. In other words, centerlines run down the middle of the road. If the road has a curve, the centerline radius will be halfway between the inner and outer radii of the road (assuming that the curves are concentric). Notice that in Roswell Street, the offsets from the centerline to the sidelines are the same on both sides, 20.00 ft. The centerline is parallel to the sidelines so both would share the same bearing.

In Fig. A.43, the dashed line running down the middle of June Street is a Centerline, but the line running through Wally Way is a baseline.

A Baseline is generally not in the middle of the road layout, nor does it necessarily run parallel to the sidelines. In Fig. A.43, Wally Way has a baseline. Notice that the baseline has an angle point at 29 + 11.03. When an angle occurs in a

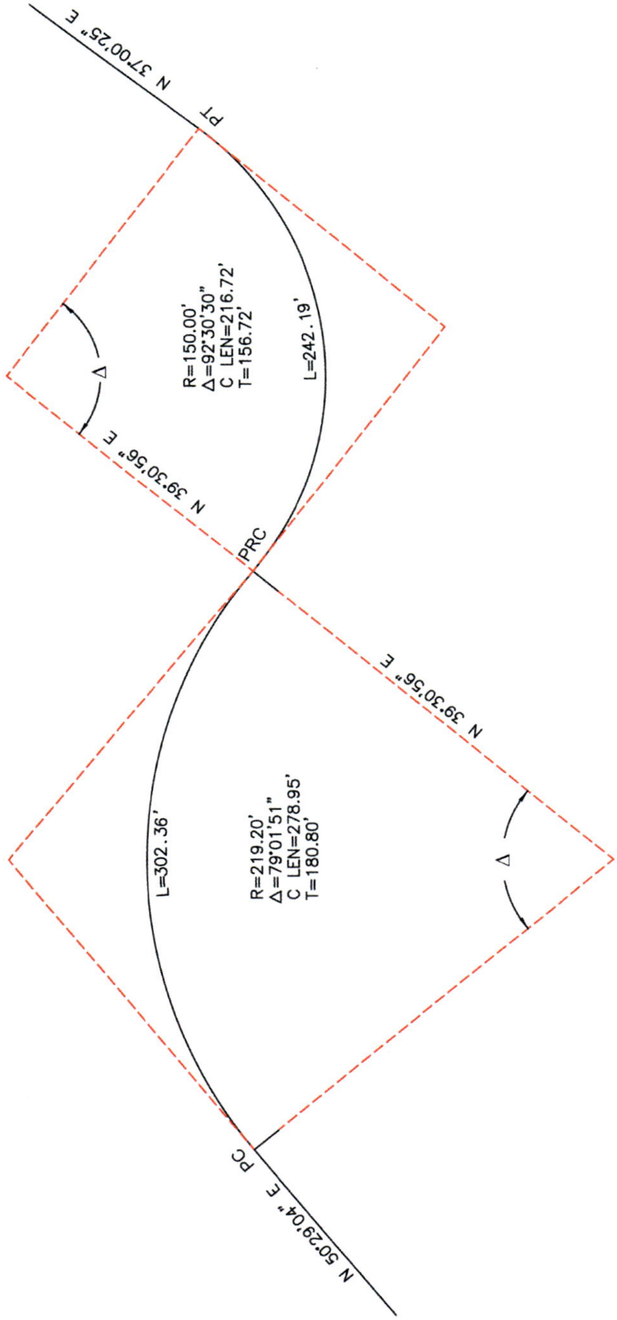

Fig. A.41 Reverse curve

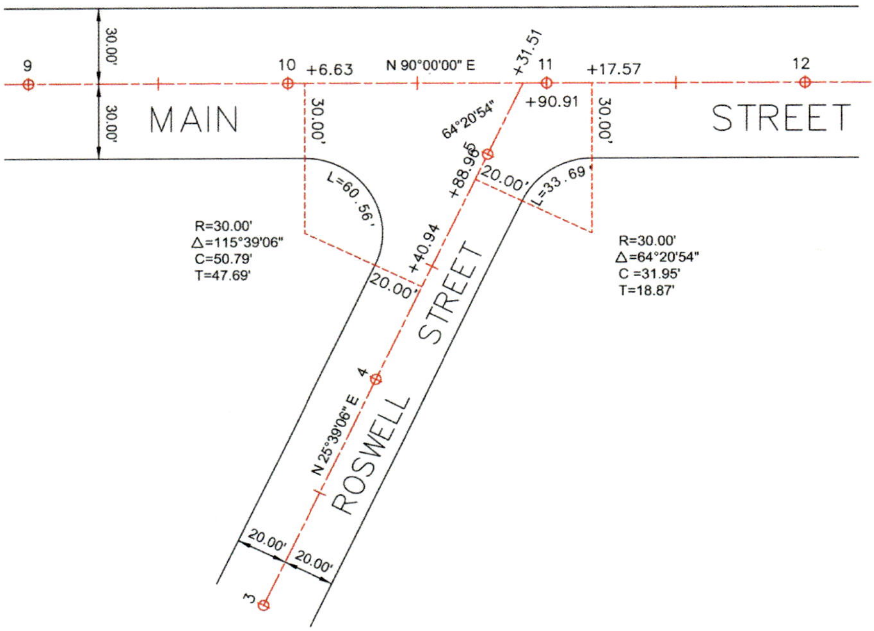

Fig. A.42 Road stations along centerline

baseline or centerline, the station numbering does not reset to zero. For example, the distance between station 30 and 29 + 11.03 is not 100 ft. it is 88.97′.

Notice that along the baseline of Wally Way, the offsets from the baseline to the curves at June Street have different dimensions. June Street has a centerline so the offsets are equal (20.00 ft.).

Both baselines and centerlines are reference lines which are measured with great precision. From a practical point of view, working along the sidelines of a road or highway can be difficult because of the many obstructions such as trees, walls and fences. It is therefore much more common to work in the relatively unobstructed area at or near the center of the roadway where baselines and centerlines are located. In roads that do not have medians, great care must be taken because of the danger from moving vehicles, so in these cases it is not uncommon to work on some convenient offset from one of the sidelines. For example, when working in an urban area where there are sidewalks, it is often convenient and relatively safe to run survey lines over the sidewalks.

Offsets are used to locate angle points in the road and at the beginning and ends of curves. When roads are designed and laid out, baselines or centerlines are used by surveyors and engineers to establish and mark the sidelines of the roads. For example, in Fig. A.43, if we wanted to set a point at the beginning of the curve at Wally Way and June Street, we could set our total station on baseline station 28 + 8.26, turn an angle of 90° from the baseline and measure over 33.66′ to set the point. This is a very important point: unless the offset line has a specific bearing and distance, it is assumed to be at 90° to the baseline or centerline.

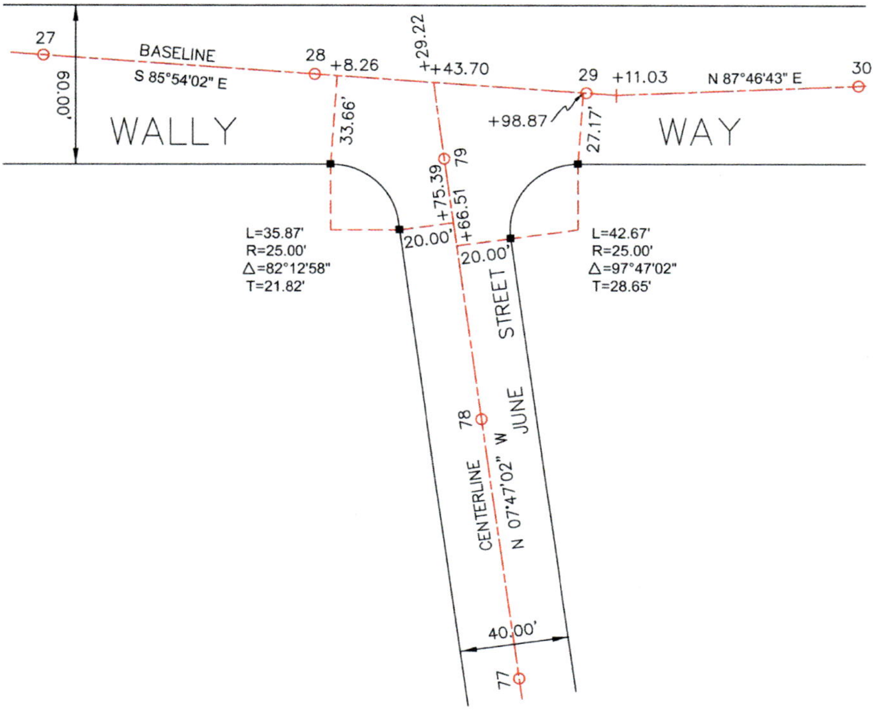

Fig. A.43 Baselines and centerlines

A.7 Calculating Area

Knowing how to calculate area is very important for land surveyors. The economic value of a parcel of land is related to its area. Some subdivision and zoning regulations require parcels of land to have minimum or maximum areas. The area of a parcel of land being conveyed is generally included in the deed description.

Like distances, which are always measured in the horizontal plane, area is also calculated in a horizontal plane. If a parcel of land is situated on a hill, the actual ground area will be larger than the area calculated on a horizontal plane. When we calculate area in boundary surveying we are generally concerned with area measured in the horizontal plane, so we will confine our discussion to calculating area in this manner.

In the U.S. area is usually measured in square feet or acres. One acre = 43.560 ft.^{2}. Often, graphic methods or planimeters were used which were not very accurate. One way to calculate area is to break up a parcel of land into a series of rectangles and triangles. The area of a rectangle is simply its width times it height.

To calculate the area of the parcel in Fig. A.44, we first notice that the all of the interior angles are 90°. If the angles were not 90°, we would have a parallelogram, and the area would be less than a rectangle. The area of a rectangle is given by:

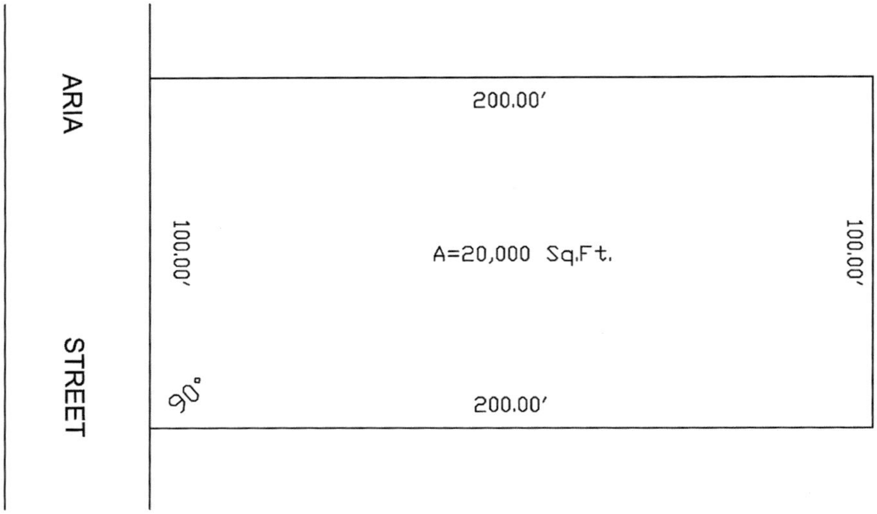

Fig. A.44 Area of a rectangular parcel

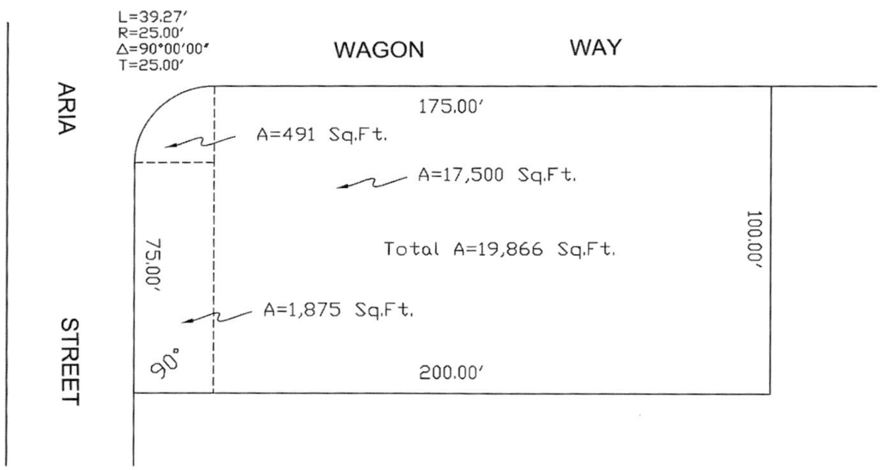

Fig. A.45 Area of corner lot

$$A = L * W = 200 * 100 = 20{,}000\,\text{ft.}^2$$

Many corner lots have curves at the intersections. One way to calculate the area of such a parcel is to break it up into rectangles or triangles. The area inside the circular sector is calculated as a portion of a circle.

Consider Fig. A.45 which shows a rectangular parcel bounded by two streets. In order to calculate the area, the parcel is broken up into 3 sections and the area of each is calculated. The area of the circular sector is calculated as a fraction of a circle. In this case the Delta is 90° so we have 1/4 of a circle. The area of a circle is given by:

$$A = \pi r^2$$

So, for a full circle with a radius of 25.00′ we calculate the area as:

$$A = \pi * 25.00^2 = 1{,}963.50\,\text{ft.}^2$$

The area of ¼ of the circle is:

$$A = 1{,}963.50 * 0.25 = 491\,\text{ft.}^2$$

If the Delta were not 90° we would simply use a ratio of the Delta to 360° as we did when we calculated the curve length in the section on road geometry. Another way to approach calculating the area of the parcel shown in Fig. A.45 would be to calculate the area of the entire parcel, as if the curve did not exist then subtract the area outside the curve from the lot area. The lot area without the curve would simply be: 100*200 = 20,000 square feet. We have already calculated the curve area of 491 square feet. The area outside the curve would be the difference between the curve area just calculated and a 25 foot square which has an area of 625 square feet (25*25 = 625). So, the area outside the curve would be: 625–491 = 134 square feet. The lot area would therefore be 20,000–134 = 19,866 square feet.

We can also use triangles to calculate area. If we know the lengths of each side of a triangle, we can use the following equations (Heron's formula) to calculate the area.

$$Area = \sqrt{s(s-a)(s-b)(s-c)} \tag{A.16}$$

$$s = \frac{1}{2}(a+b+c) \tag{A.17}$$

where a, b and c are the lengths of the sides of the triangle.

Let us use the parcel of land shown in Fig. A.46 to see how we can use triangles to calculate the area of a parcel.

This is the same parcel of land that is shown in Fig. A.21 with some of the information removed for clarity. We break the parcel into two triangles by inversing between the opposite lot corners. We now have two triangles so we can use Eqs. A.16 and A.17 to calculate the areas.

Triangle 1:

$$s = \frac{1}{2}(a+b+c) = \frac{1}{2}(150.72 + 196.34 + 207.83) = 277.445$$

$$A = \sqrt{277.445(277.445-150.72)(277.445-196.34)(277.445-207.83)} = 14{,}089.5\,\text{feet}^2$$

The same procedure would be used to calculate the area of the second triangle, and the two would be added together.

The parcel shown in Fig. A.47 is an example of how the area of an irregularly shaped parcel can be calculated using triangles. As you can see from this exercise,

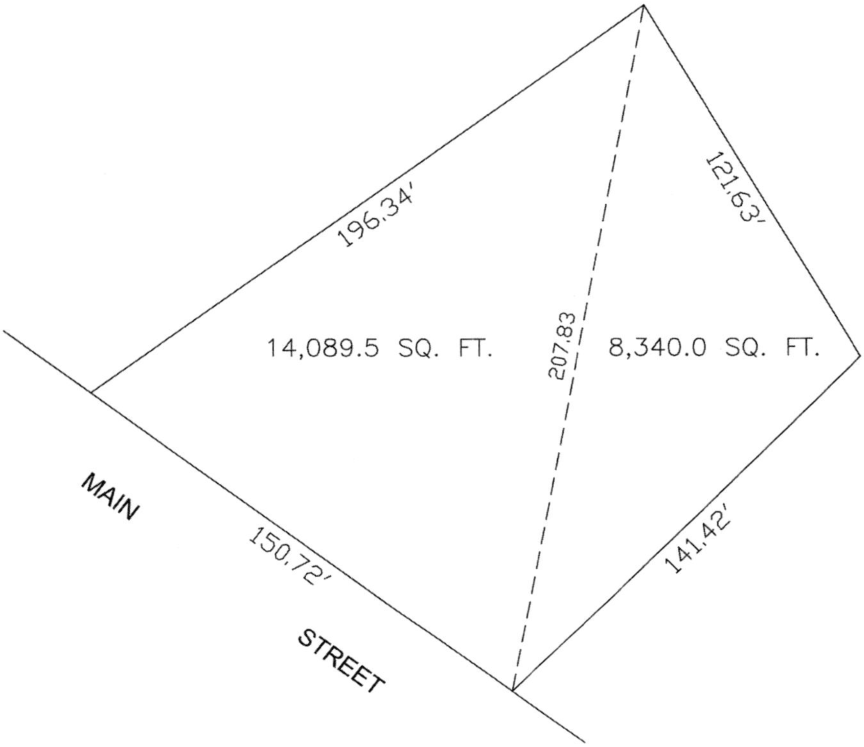

Fig. A.46 Using triangles to calculate area

calculating area this way is a lengthy and time consuming process. Because of the large number of calculations, there is also a greater possibility of making a mistake.

A very common way to calculate area is based on double meridian distances or coordinates. When latitude and departures and coordinates are calculated in tabular format as in the section on adjusting traverses, these methods are convenient

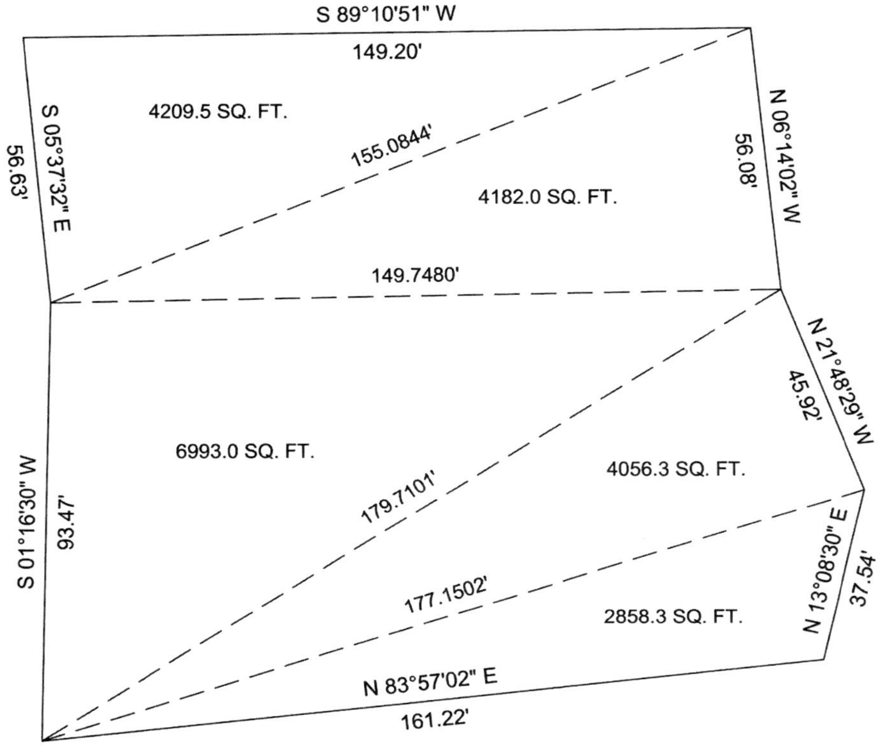

Fig. A.47 Area of irregular parcel using triangles

because they use this same information to calculate area. However, neither of these methods are used very much by surveyors these days, except when they are a part of computerized survey calculation software. Thankfully, the laborious methods of performing area calculations are now performed by software in seconds rather than the hours it took using manual calculation methods.

Glossary of Terms Used in Boundary Surveying

Accretion The gradual accumulation of land by natural causes along a water boundary

Accuracy The amount of deviation from the true value of a measurement

Acquiescence When a land owner impliedly accepts a boundary location established by an abutter

Actual Possession Possession which consists of actual physical occupation of property

Adjoiner Abutting property

Aliquot Part In the PLSS, a standard division such as a half section, quarter section, half-quarter section, etc

Angle A figure formed by the intersection of two lines, and the measurement of the degree of separation between the two lines

Apportionment The distribution of excess or deficiency in a number of parcels which were all created simultaneously

Area The quantity of land contained within the boundaries of a parcel

Artificial Monuments A manmade marker such as a concrete bound, or drill hole

Avulsion The sudden removal or addition of land by the action of water

Azimuth The direction of a line relative to north (sometimes relative to south)

Backsight A target set up over a point where the instrument will set zero

Base Line In the PLSS, a line of latitude running through an initial point. Also, a random line used in a road layout

Bearing The direction of a line between 0° and 90°, measured clockwise or counterclockwise from north or south

P. Gay, *Practical Boundary Surveying*, DOI 10.1007/978-3-319-07158-9

Blaze A mark or marks on a tree trunk for marking a survey point or boundary point or line

Boundary See property line

Break Tape To make a horizontal measurement, usually while measuring up or down a slope, by using less than the full length of a tape

Called-For A monument or other evidence called out in a conveyance

Chain A distance of 66 ft.

Chain of Title A sequence of unbroken conveyances from the original title holder to the present title holder

Chord or C The straight line distance between the PC and PT of a curve

Circle of Uncertainty A circle which has a radius equal to the maximum probable error

Closing Corner In the PLSS, it is the intersection of a meridional line with a parallel between fixed corners

Constructive Notice Notice that is implied whether or not it actually exists

Continuous When used in relation to adverse possession, the possession must be unbroken

Convergence Where meridians become closer together as they progress northerly and southerly from the equator

Coordinate System A grid where the axes are 90° apart. Also called a Cartesian Coordinate System

Corner The location of intersecting boundary lines

Curve Radius The distance between the curve and the center of the circle

Declination The angular difference between true north and magnetic north

Deed A writing signed by the grantor conveying title from grantor to grantee

Delta The angle or deflection of a curve

Differential Positioning When used with GPS, a method whereby positional correctional data is transmitted from a GPS receiver at a fixed and known station to a rover such that the rover can calculate a more accurate position than it would be able to without the correction information

Dominant Estate or Dominant Tenement The land which an easement appurtenant attaches to

Double Center To set a point on a projection of a line by sighting the point with the telescope erect and inverted and splitting the difference between the two points

Double Proportionate Measurement In the PLSS, restoration of a lost corner by proportionate measurements in both east–west and north–south directions from existing monuments

Drill Hole A hole made in stone or concrete for the purpose of locating a point or property corner

Easement The right to use the land of another person

Easement Appurtenant An easement which attaches to the land not to an individual person

Easement by Implication When a conveyance does not explicitly call for the easement but the easement must have been intended

Easement Implied from Necessity When a grantor creates and conveys a landlocked parcel of land, an easement is implied over the grantor's remaining land to a road

Easement Implied from Prior Use When a grantor creates and conveys a landlocked parcel of land and there was visible access such as a way leading over the grantor's remaining land to a road, an easement is implied

Easement Implied from a Plat Common in subdivisions containing private roads, it addresses the rights which owners in a subdivision have over roads not necessarily required to access a public road

Easement in Gross An easement which is owned by a person and does not attach to the land

EDM Electronic Distance Measuring Device

Electronic Distance Measuring Device A device, usually attached to a total station, which uses light beams to precisely measure distance

Electronic Data Collector A small digital computer with a screen which connects to a total station or GPS receiver for the purpose of collecting data from the device. If the total station is servo driven the data collector can drive the instrument

Error of Closure The length or length and direction of the gap in a geometric figure where the end point is intended to be the same as the beginning point

Estoppel When a property owner induces an abutter to act in a way that is likely to damage the abutter's interests. Detrimental reliance

Exclusive When used in relation to adverse possession, the possessor must not share the possession with the owner

Express Easement A written easement

Fee An ownership interest in real property

Fee Simple Absolute Absolute and exclusive ownership of land

Field Book A small bound or loose leaf book used to record survey field data

First Guide Meridian East and West Meridians established 24 miles east and west of the principal meridian

Foresight A target or prism to which an instrument will measure an angle and/or angle and distance

Give Line The process by which a person holds a prism pole or plumb bob string over a point so it can be sighted by the instrument

Global Navigation Satellite Systems (GNSS) A constellation of satellites and ground stations owned by the United States and Russia used for the positioning of receivers on the earth's surface

Global Positioning Systems (GPS) A constellation of satellites and ground stations owned by the United States used for the positioning of receivers on the earth's surface

Grantor One who conveys land or some interest in land

Grantee One who acquires land or some interest in land

Grid North The direction of the north–south grid lines on a state plane coordinate system

Gun A common name for a transit, theodolite or total station

Horizontal Measurement or Horizontal Distance A measurement made in a plane which is parallel to the horizon

Hostile When used in context with adverse possession and prescription, the possession must be adverse to the interests of the owner

Incorporate To make a part of

Initial Point In the PLSS, a fixed point from which townships are established

Instrument A surveying instrument such as a transit, theodolite or total station

Instrument Person The person charged with the responsibility for the care and use of an instrument such as a level, transit, theodolite or total station

Intention of the Parties to the Conveyance What the grantor and grantee agreed to at the time of the conveyance

Intervisible The ability to visually see between points

Junior and Senior Rights Rights in property acquired through a sequence of conveyances by a common grantor

L or LC The length of a curve

Latitude and Departure Latitude is the distance measured northerly on a coordinate grid. Departure is the distance measured easterly on a coordinate grid

Legal Description The portion of a deed which describes the property or interest in property being conveyed

Line The direction in which the instrument vertical cross hair is pointing

Link 1/100 of a Chain (0.66 ft.)

Littoral Rights Property rights related to water boundaries along an ocean or lake

Locate Using an instrument to turn an angle and measure a distance to a target or other object

Locus The property at issue. The property being surveyed

Lost Corner In the PLSS, where the location of the corner cannot be determined from marks or evidence, so that it must be replaced by measurements from other corners

Low Distortion Projection A two dimensional rectangular coordinate system located at an elevation designed to minimize distortion between measurements made on the ground and those on the coordinate system

Magnetic Dip The angular downward tilt of a compass needle as measured from horizontal

Magnetic North The direction in which the magnetic needle of a compass points

Mean High Water The average of all high water heights observed over a 19 year period

Meander Corner In the PLSS, an intersection of a meander line with a section or township line

Meander Line In the PLSS, a series of traverse lines at mean high water

Measure Angle To use an instrument to measure an angle

Meridians North–south lines converging at the north and south poles

Metes and Bounds A method of describing the limits or boundaries of a parcel of land by distance and direction and by reference to physical monuments

Monument A physical object used to fix the location of a property corner or line

Natural Monument A naturally occurring boundary marker such as a boulder, stream or tree

Obliterated Monument In the PLSS, where there is no remaining evidence of a monument but there is evidence of its location

Occupied Point A point on which the instrument or GPS receiver is set

On Line A condition whereby a target, prism pole, plumb bob, etc. is in line with the vertical cross hair of an instrument

Open and Notorious When related to adverse possession and prescription, the possession must be so openly visible that it would be obvious to the owner

Optical Plummet An optical device including a right angle prism which allows a tribrach to be precisely centered over a point

Original Survey A survey in which new boundaries are created

Overburdening an Easement Use of an easement which was not contemplated when the easement was created

Overloading an Easement Use of an easement to service land which was not contemplated when the easement was created

Parole Evidence Spoken Testimony

Party Chief or Crew Chief The person in charge of a survey crew

Personal Property Anything subject to ownership that is not real property

PC Point of Curvature. The point where a curve begins

Pile of Stones A heap of stones used to mark a property corner or sometimes placed at intervals along property lines

Pin A small metal cylinder usually of copper, brass or aluminum which is expanded into a hole in stone or concrete for marking a property corner

Plan, Survey Plan, Plat or Map A drawing showing the results of a survey

Point on Line A point set on an existing line

Positional Dilution of Precision (PDOP) In GPS or GNSS, a number indicating the effect of satellite geometry on the accuracy of measurements

Physical Evidence Also called Real Evidence. Physical objects such as monuments, roads, lakes and rivers

Planimeter A mechanical instrument used to trace the boundary of a plane figure for the purpose of calculating its area

PLSS The Public Land Survey System. A rectangular system of dividing land used in the western United States

Point of Beginning Used in deed descriptions to describe the point at which a meets and bounds description begins

Post-Processing When used in conjunction with static GPS, it is a software method used to adjust and calculate the coordinates of points which were previously occupied by GPS receivers

Practical Location Where the location of a boundary is not determined from a writing but by agreement, acquiescence, or occupation

Precision The deviation in a series of repeated measurements to the same point or between the same points

Prescription Acquiring an easement through use, not by virtue of a writing

Principal Meridian In the PLSS, a line of longitude running through an initial point

Prism A glass reflector also called a corner cube prism, which is used to reflect light from an EDM back to the EDM so that the distance between the EDM and the Prism can be measured

Prism Offset An adjustment to the distance measured between an EDM and a prism used to account for the speed and path of light and the mounting of the prism

Prism Pole A portable extendable pole used for mounting a prism commonly used for measuring to physical evidence

Private Street A street which may be used only by those having deeded rights

Privity A mutual or successive relation between parties to the same real property

Property Line The limit of ownership of land held by title deed or by marks of enclosure if possession is without title deed. It is also called a boundary

PT Point of Tangency. The point where a curve ends

Public Street A street over which the general public has a right of access

Quadrant One quarter of a circle: 90°. Used in surveying to refer to the four quadrants: Northeast, Southeast, Southwest and Northwest

Rangefinding Optically measuring distance using stadia hairs in a surveying instrument telescope

Real Property Land and things affixed to the land

Rebar Steel concrete reinforcing bar sometimes placed into the ground as a means of locating a property corner

Record Boundaries Boundaries described in a deed or plan

Record Evidence Written evidence recorded with a public agency such as deeds and plans

Record Monuments A boundary marker called for in a deed such as the name of an abutter

Reference Station When used with RTK GPS, a GPS receiver located at a fixed and known point which is capable of transmitting correction data to a GPS rover

Reflectorless Total Station A total station which is able to measure distances to an object without the need for using a prism

Reliction The increase in land by permanent withdrawal of water along a water boundary

Reserved Easement An easement created by a grantor that reserves a right to use property of the grantee

Responsible Charge Direct control and supervision of a person working under a licensed land surveyor

Reticle Horizontal and vertical cross hairs in a surveying instrument telescope used to accurately sight a target

Retracement Survey A resurvey of an original survey

Riparian Rights Property Rights along a river or stream

Road A way used for travel. Same as Street

Robotic Total Station A total station containing servo motors which are capable of positioning the horizontal and vertical axes in order to measure to evidence and transmit data to a remote data collector

Rod A distance of 16.5 ft.

Rod Person Usually, the least experienced crew member. Often, the person who uses a prism pole to locate evidence

Rover A GPS receiver located on a prism pole or tripod which is moved around from point to point for the purpose of locating evidence or setting points. It normally receives correction data from a reference station

Scope of an Easement The type of use allowed by an easement

Section In the PLSS, a portion of a township approximately one mile square containing 640 acres

Servient Estate or Servient Tenement The land burdened by an easement appurtenant

Set an Angle To set a specific angle on an instrument

Set Up The process of positioning and leveling a tripod or prism pole and bipod over a point

Sexigesimal System A system based on 60, where each degree contains 60 minutes and each minute contains 60 seconds

Shot A term describing the procedure of an instrument measuring an angle and a distance to a target or other object

Sideshots Leaving zero set on a backsight and taking shots on multiple targets

Single Proportionate Measurement In the PLSS, restoring a lost corner by using proportionate measurements along a line from other recovered monuments

Stadia Hairs Short horizontal cross hairs in a surveying instrument telescope precisely positioned above and below the horizontal cross hair used for rangefinding

Standard Parallels In the PLSS, parallels established 24 miles north and south of base lines

State Plane Coordinate System One or more two dimensional rectangular grids established by a state for locating points

Statute of Frauds A law requiring that the conveyance of interests in land by evidenced by a writing signed by the grantor

Street A way used for travel. Same as Road

Street Line The sideline of a street or way

Survey Foot 1,200/3,937 miles

Tacking When used in relation to adverse possession and prescription, the occupation by predecessors in title can be used, along with the possession of the present occupier to satisfy the time period established by the statute of limitations

Taking a Shot When a total station is used to measure the angle and distance to a point

Tangent A line that touches a curve such that it is perpendicular to a radius line of the curve. Also, an extension of a street line from the PC or PT of a curve

Tangent Screws Fine positioning screws on a surveying instrument horizontal and/or vertical axis

Target A device usually containing a prism and/or highly visible markings which is set over a point and which enables an instrument to sight and measure to the point

Temperature Adjustment A correction applied to a distance made using a steel tape and based on the difference between ambient temperature and 68 °F

Thalweg The deepest part of a stream

Theodolite A mechanical and optical device having graduated glass circles which can measure horizontal and vertical angles. A theodolite may have optical encoders used to read the circles

Thread The middle of a stream

Tie Line A dimensioned survey line, usually not a property line, used to connect boundaries. Tie lines are often used along an irregular and roughly dimensioned water boundary as a means of connecting the accurately described points of a boundary

Torrens System A system of title registration, usually managed by the state

Total Station An electronic digital surveying instrument used to measure horizontal and vertical angles and distances, with provision for electronic storage of data and/or transmission of data to external devices

Township In the PLSS, a quadrangle having sides approximately 6 miles in length which correspond to the parallels and meridians

Tracking Total Station A total station which is capable of locking onto a prism or other device and tracking it as it moves

Transit A mechanical and optical device used to measure horizontal and vertical angles

Traverse A series of randomly located survey lines run in order to locate evidence and set property corners

Trespass The intentional intrusion or invasion of the land of another person

Tribrach A fixture attached to the top of a tripod which can be leveled and positioned and used to attach surveying instruments and other measuring devices to the tripod

Tripod A fixture having three, usually adjustable, legs used to support surveying instruments and other measuring devices

True North A direction parallel to the earth's axis which points toward the north pole

Turn Angle To use an instrument to measure or set an angle

Unwritten Title Ownership of property that is not evidenced by written title

Variable Width Street A street which does not have a fixed width

Variation A measure of how magnetic north varies with time

Vernier A specially divided scale used to increase the precision of reading the horizontal or vertical circle of a surveying instrument

Wiggle In To make a series of adjustments to the occupied point so that it eventually becomes located at the desired location

Witness Corner In the PLSS, a point on a survey line near a corner used when it is not practical to monument the corner itself

Written Evidence Documents such as deeds and plans

Zero Instrument The process of setting an instrument's zero on a target

Further Readings

[ACS89] American Congress on Surveying and Mapping and the American Society of Civil Engineers. (1989). *Definitions of surveying and associated terms.*

[Bre42] Breed, C. B. (1942). *Surveying*. New York: Wiley.

[BrH51] Breed, C. B., & Hosmer, B. L. (1951). *The principles and practice of surveying*. New York: Wiley.

[BrH26] Breed, C. B., & Hosmer, G. L. (1926). *Principles and practice of surveying: Volume II, higher surveying*. New York: Wiley.

[BrW84] Brinker, R. C., & Wolf, P. R. (1984). *Elementary surveying*. New York: Harper & Row Publishers.

[Bro69] Brown, C. M. (1969). *Boundary control and legal principles*. New York: Wiley.

[BRW81] Brown, C. M., Robillard, W. G., & Wilson, D. A. (1981). *Evidence and procedures for boundary location*. New York: Wiley.

[BLM09] Bureau of Land Management, United States Department of the Interior. (2009). *Manual of surveying instructions*. Washington, DC: Government Printing Office.

[CSW93] Cunningham, R. A., Stoebuck, W. B., & Whitman, D. A. (1993). *The law of property* (2nd ed.). St. Paul, MN: West Publishing Co.

[DaF40] Davis, R., & Foote, F. S. (1940). *Surveying theory and practice*. New York: McGraw-Hill Book Co.

[FGC98] Federal Geodetic Control Subcommittee, Federal Geographic Data Committee. (1998). *Geospatial positioning accuracy standards.*

[Har90] Harbin, L. H. (1990). *Land surveyor reference manual*. San Carlos, CA: Professional Publications Inc.

[KaB84] Kavanaugh, B. H., & Bird, G. (1984). *Surveying: Principles and applications*. Reston, Virginia: Reston Publishing Company Inc.

[Lon02] Dr. Londe, M. D. (2002). *Bureau of land management, standards and guidelines for cadastral surveys using global positioning methods.*

[Mad84] Madson II, T. S. (1984). *The mathematics of land surveying; land surveyor's seminar*. Gainesville, Florida: Land Surveyor's Seminar.

[McC76] McCormac, J. C. (1976). *Surveying*. New York: Prentice-Hall.

[MiS77] Mitchell, H. C., & Simmons, L. G. (1977). *The State coordinate systems (a manual for surveyors)*. U.S. Department of Commerce, Coast and Geodetic Survey, Special Publication No 235.

[Mof75] Moffitt, F. H. (1975). *Surveying* (6th ed.). New York: Intext Educational Publishers.

[Rab74] Rabin, E. H. (1974). *Fundamentals of modern real property law*. Mineola, NY: The Foundation Press Inc.

P. Gay, *Practical Boundary Surveying*, DOI 10.1007/978-3-319-07158-9

[Ray39] Rayner, W. H. (1939). *Elementary surveying*. New York: Van Nostrand Co.
[Sie79] Siedel, G. J. (1979). *Real estate law*. Eagan, MN: West Publishing Company.
[VeR97] Vernon, R. C., & Robert, C. B. (1997). *Professional surveyor's manual*. New York: McGraw Hill.
[WSD04] Washington State Department of Natural Resoruces. (2004). *Standards and guidelines for land surveying using global positioning system methods.*

Index

P. Gay, *Practical Boundary Surveying*, DOI 10.1007/978-3-319-07158-9

Printed by Printforce, the Netherlands